To Beth

Thanks for attending; Enjoy the story...

# NESIRITIDE

Also by Roger M. Mills, MD

*Clinical Management of Heart Failure* (with James B. Young, MD)

*Practical Approaches to the Treatment of Heart Failure* (with James B. Young, MD), first edition, 2001; second edition, 2004

# NESIRITIDE

*The Rise and Fall of Scios*

ROGER M. MILLS, MD

**NESIRITIDE**
**THE RISE AND FALL OF SCIOS**



*iUniverse books may be ordered through booksellers or by contacting:*

*iUniverse*
*1663 Liberty Drive*
*Bloomington, IN 47403*
*www.iuniverse.com*
*1-800-Authors (1-800-288-4677)*



*ISBN: 978-1-4917-9764-8 (sc)*
*ISBN: 978-1-4917-9763-1 (hc)*
*ISBN: 978-1-4917-9762-4 (e)*

*Library of Congress Control Number: 2016908430*

*Print information available on the last page.*

*iUniverse rev. date: 06/28/2016*

This book is dedicated to Katherine, who patiently lived through the story, and to Cody and Shadow (a.k.a. the Beau Zeau), who listened to it on our walks.

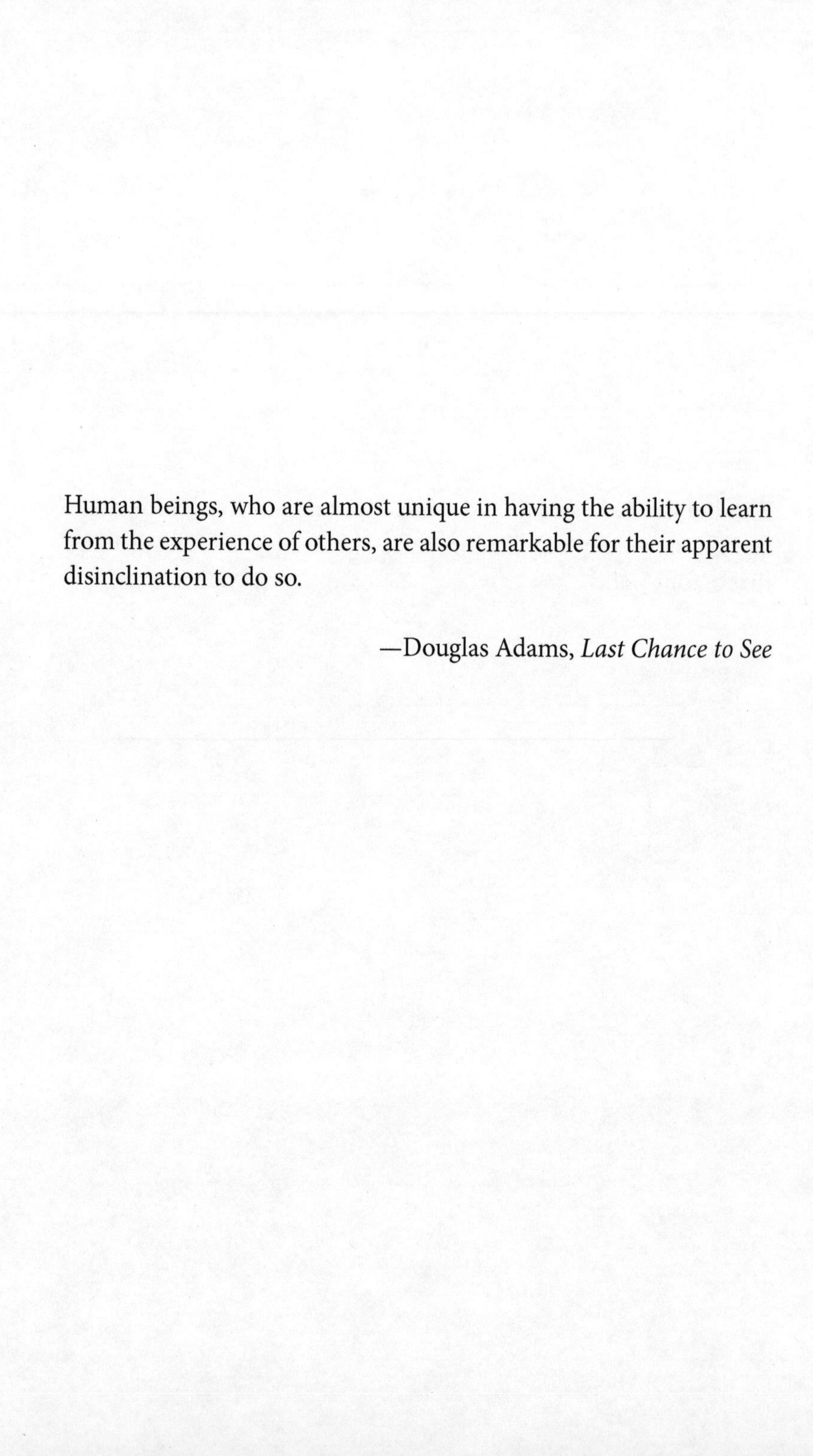

Human beings, who are almost unique in having the ability to learn from the experience of others, are also remarkable for their apparent disinclination to do so.

—Douglas Adams, *Last Chance to See*

# Contents

# *Preface*

When I had the chance to make a late-life career change from academic cardiology to the pharmaceutical industry, I had no idea what kind of adventures the next decade would provide. As it turned out, I would experience the rise and fall of a new drug for acute heart failure, nesiritide (brand name: Natrecor), and of Scios, the biotechnology company that brought it to market. I had participated in the early stages of the rise as a clinical investigator in the phase 2 trial and as an academic consultant to the company. I witnessed the fall firsthand as an insider from the position of vice president for medical affairs at Scios.

Along the way, ordinary day-to-day events were often colored with drama. They could have been lead-ins for the day's episode of an afternoon soap opera. After the FDA rejected Scios's 1999 application for marketing approval, would Dick Brewer pull his small independent company together and gain an approval on a second try? After the hugely successful launch of Natrecor in 2003, would management remember the warnings it had heard from the FDA and the critics in academic medicine? Could a more objective assessment of the risks and benefits of nesiritide have saved the drug and the company?

As the events in this book played out, I promised myself that in retirement I would write the story as objectively as I could. This book is not a "sour grapes" fable. As you will see, there were enough mistakes on all sides of the nesiritide controversy to go around.

I've always admired Jack Webb's character, Sergeant Joe Friday, from the TV series *Dragnet*. For many years, I kept a picture of Webb with his trademark line, "Just the facts," on my desk.[1] It reminded me to stay focused. Sgt. Friday insisted on understanding the facts, not the embellished stories of biased witnesses. The relentless focus that Joe showed in uncovering the facts provided the drama of the *Dragnet* stories. I hope that the facts in *Nesiritide* will give you a sense of the drama of the rise and fall of a small California biotech company and its product.

I studied nesiritide in clinical trials, and I used it in clinical practice. I felt that it was a safe and effective short-term option for managing appropriate patients in appropriate circumstances. Otherwise, I would not have taken a job with Scios. As the clinical trial data have repeatedly confirmed, heart failure patients receiving nesiritide had outcomes similar to those treated with intravenous nitroglycerin or nitroprusside, while requiring much less intensive (and expensive) care. On the other hand, there was no evidence to support the use of nesiritide for regularly scheduled outpatient infusions, and I did not condone that practice. During the public controversy about nesiritide, I felt that some of the tactics the drug's critics employed went beyond the pale, and I still do about some specific events. But like Joe Friday, I firmly believe that the facts will tell the story best.

Writing this story has expanded my viewpoint and markedly altered my opinions. There are several themes that are woven together in the story:

- the cumulative nature of scientific effort over time and across space
- the powerful impact of public media on even the largest companies
- the rapidly changing nature of drug safety efforts
- the organizational challenges of success

---

[1] Apparently, Webb never used that phrase during the show (*Wikipedia* 2016).

The scientific story relating the discovery and characterization of nesiritide shows the time that working out the science requires and the continual importance of building on the discoveries of others. I truly enjoyed writing this section.

The critics of Natrecor, no matter how complex and varied their motivations, used the media skillfully. They felt that achieving their ends justified the means. They deserve an acknowledgment that their tactics worked. Using the power of the press, they forced a major pharmaceutical company to do a multimillion-dollar safety trial with a drug that was commercially dead in the water. Even if I can't agree with how they did it, I have come to understand that their sincere concerns and their earlier warnings had been ignored.

I felt that drug safety,[2] which is a major element of the nesiritide story, warranted some additional discussion. I've added my perspectives gained from chairing J&J's company-wide Cardiovascular Safety Group at the end of the story. I believe that changes in the FDA's authority to mandate postmarketing studies and improvements in the evaluation of drug safety have relegated some of the lessons from the Scios experience to the status of historical interest only.

Success brings its own challenges. The Scios story is a cautionary tale. I hope that entrepreneurs embarking on new ventures can learn from the difficulties Scios and J&J encountered in attempting to transition the Scios organization from a small biotech to a national-level pharmaceutical company.

Without the help and friendship of Robert Hobbs of the Cleveland Clinic and Darlene Horton at Scios, I would never have had opportunity to be involved with the phase 2 development of nesiritide. The success of that early involvement rested on the efforts of two absolutely outstanding nurses, Kay Worley Price and Dana Leach, who did the actual research work at Shands Hospital at the University of Florida.

I want to thank Scios management under J&J, particularly Jim Mitchell and Randall Kaye, for extending me the opportunity to join

---

[2] The newly popular buzzword for drug safety is *pharmacovigilance.*

the company, and Jim Barr for patiently, if gruffly, teaching me the essential facts of business life.

Chris Ernst handled public relations for Scios during the press crisis. She maintained a chronological file of press releases and other key publications that included many of the source documents I have cited. I deeply appreciate her thoughtfulness in sharing a copy of her file with me. Thank you, Chris.

Finally, I cannot adequately acknowledge the support of my wife, Katherine. She put up with the corporate moves and provided a sounding board for all my worries.

Roger M. Mills, MD
Dexter, Michigan
January 2015

# *Introduction*

This is a true story, and for some readers, the specialized vocabulary I've had to use will be unfamiliar. The first time that I have used a term with special meaning in the language of the pharmaceutical industry, of statistics, or of medicine, I have explained the term first and then put it in italics before using it subsequently without explanation.

Throughout the story, most of the characters are affiliated with various organizations. The major ones, including "academic medicine," the Cleveland Clinic, the Food and Drug Administration, and Johnson & Johnson, all deserve a brief introduction.

The term *academic medicine* broadly defines a large group of physicians in the United States whose professional lives include various combinations of medical practice, teaching, and research. These academic physicians have a formal contractual relationship and an academic appointment with a university school of medicine. In contrast, the term *private practice community* refers to physicians who provide health care services in a variety of settings: private solo practice, specialty or multi-specialty groups, or, increasingly, in hospital-owned group practices. Their common denominator is that they have little to no involvement in teaching and research and no formal academic appointments.

Many people know the Cleveland Clinic—where several of the important characters in this story have worked—as a major cardiovascular center, but it is also an academic center with research, clinical training programs, and the Cleveland Clinic Lerner College of Medicine (Cleveland Clinic 2016). The clinic has a long history of important clinical research in the field of heart disease, including the development of coronary angiography and coronary artery bypass graft surgery (Sheldon 2008).

The United States Food and Drug Administration (FDA) has a very broad mission;[3] in this story, the narrative focuses on encounters between Scios Inc., a small California biotechnology company developing a cardiovascular drug, and a single section of the FDA, the Cardio-Renal Division of the Center for Drug Evaluation and Research. Dr. Raymond J. Lipicky was the director of the cardiorenal division from March of 1982 to February of 2002. Dr. Robert Temple serves as the FDA deputy director for clinical science in the Center for Drug Evaluation and Research (CDER) and is also acting deputy director of the Office of Drug Evaluation.

Johnson & Johnson (J&J) is a major diversified US health care corporation. Although many people associate the J&J name with Band-Aids and baby powder, J&J is a very large company with three major areas of business activity: consumer products, medical devices and diagnostics, and pharmaceuticals. The J&J corporate culture emphasizes an internal code of ethical behavior called "the Credo" (Johnson & Johnson 2016). Readers will see as this story develops that when reports in the medical literature and headlines in the press called the safety of a J&J product into question, given its corporate values, it was unthinkable that J&J would not respond with an appropriate study.

---

[3] With regard to the Food and Drug Administration (FDA), interested readers may want to review a brief history of the FDA available on the agency's website (US FDA 2015a). I also recommend Lawrence Friedhoff's book, *New Drugs: An Insider's Guide to the FDA's New Drug Approval Process* (Friedhoff 2009) for readers who want to learn more about how drugs are evaluated and approved in the United States.

The process of drug discovery and development follows a long, hard trail. When a compound with therapeutic potential is discovered, extensive preclinical testing in animals is required to understand its pharmacology and toxicology. If the preclinical profile looks reasonable, then the three-phase process of clinical testing begins. Clinical testing takes time and a reliable long-term supply of money. Phase 1 testing involves giving a wide range of doses to healthy human subjects to assess the drug levels associated with increasing doses and to document how humans absorb and excrete the drug. Then comes phase 2 testing, the first administration of the drug to human patients with the target disease. Phase 2 allows the clinical research and development team members their first look at whether or not the drug might really be effective. Finally, if the compound advances into phase 3 trials, the drug will be compared to an appropriate control, either an inert placebo or an "active control" drug with some known effectiveness in one or more large randomized blinded controlled trials. In the United States, after successful phase 3 testing, the drug developer must bring all the data to the FDA (often called "the agency") and request marketing approval. When the agency feels that a compound is novel or controversial, the review and approval process often includes an advisory committee ("Ad Comm") meeting. Advisory committees are made up of external experts with special knowledge in the relevant therapeutic area.

Also for the nonmedical reader, I have included two illustrations here in the introduction. Both of the figures are reiterations of diagrams that I have drawn on blackboards (or whiteboards) countless times in conferences for medical residents and cardiology fellows.

Figure 1 shows the answer to one of my favorite questions for students, "How many hearts and how many lungs does a normal human have?" Anatomically, of course, the answer is, "One heart and two lungs." But much of this book is about physiology, the science of how our bodies work, and physiologically the correct answer is, "Two hearts and one lung."

Normal adults have two circulations. There is a low-pressure, low-resistance, volume-adapted *pulmonary* circulation, and a high-pressure, variable-resistance *systemic* circulation. The right heart supplies the lungs via the pulmonary circulation. With exercise, the pulmonary circulation allows large volumes of blood to flow through the lungs at low pressure, exchanging waste carbon dioxide ($CO_2$) for fresh oxygen ($0_2$) with only a small amount of additional work for the right heart. In contrast, the left heart supplies the systemic circulation at high pressure. This allows a complex system of variable resistances throughout the body to control precisely where the blood flows. With exercise, systemic pressure rises and the left heart works harder, generating higher pressure with an increased heart rate; however, resistance to blood flow in the exercising muscles drops. This fall in local resistance allows large volumes of freshly oxygenated blood to flow to the muscles.

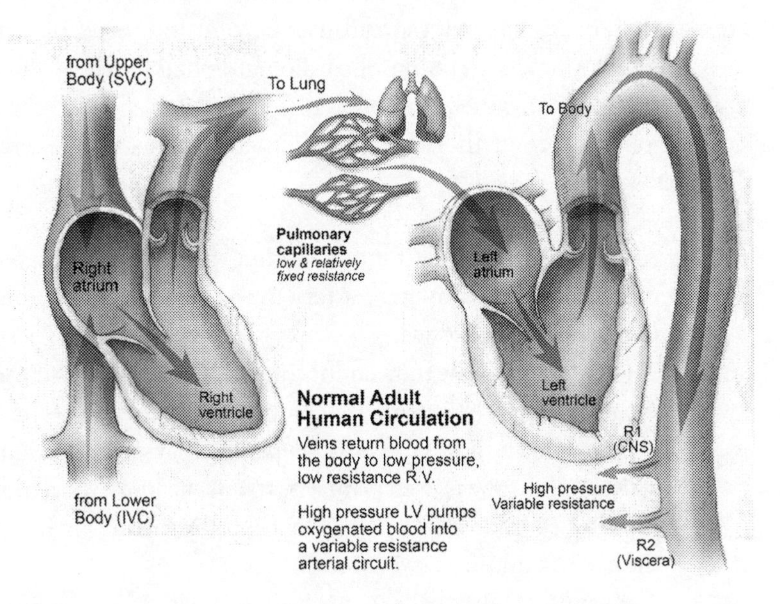
from Upper
Body (SVC)
To Lung
To Body
Pulmonary
capillaries
low & relatively
fixed resistance
Right
atrium
Left
atrium
Right
ventricle
Left
ventricle
Normal Adult
Human Circulation
Veins return blood from
the body to low pressure,
low resistance R.V.
High pressure LV pumps
oxygenated blood into
a variable resistance
arterial circuit.
R1
(CNS)
High pressure
Variable resistance
R2
(Viscera)
from Lower
Body (IVC)

Figure 2 shows a Swan-Ganz flow-directed balloon right heart catheter in place in the pulmonary circulation. In Chapter 4, I have described the development and use of the catheter.

The Swan-Ganz catheter was a relatively safe and inexpensive tool that moved cardiac physiology from the textbooks and the research laboratories to the bedside. In a way, it did for the management of critically ill heart failure patients what GPS guidance did for flying on instruments. Cardiologists could define a patient's circulatory status, adjust course with changes in medications or fluids, and then see the results in real time. They were no longer flying blind. One of the consequences of its widespread use, beyond improved patient care, was that a new generation of cardiologists had a much better understanding of how the heart worked.

As you can see in the figure, the catheter consists of several different systems:

- A fluid-filled hollow tube extending from the tip of the catheter in the pulmonary artery to the bedside connection to a pressure transducer.
- An air-filled hollow tube connecting to the balloon near the tip, allowing inflation and deflation of the balloon.
- A fluid-filled hollow tube with an opening that sits in the right atrium. This tube is used primarily to inject a known amount (usually 10 ml.) of iced saline solution for measurement of the cardiac output.
- A thermistor wire that extends from just behind the balloon to a bedside coupler connecting to a dedicated computer. The computer calculates cardiac output based on the pattern of temperature changes in the blood after the iced saline injection.

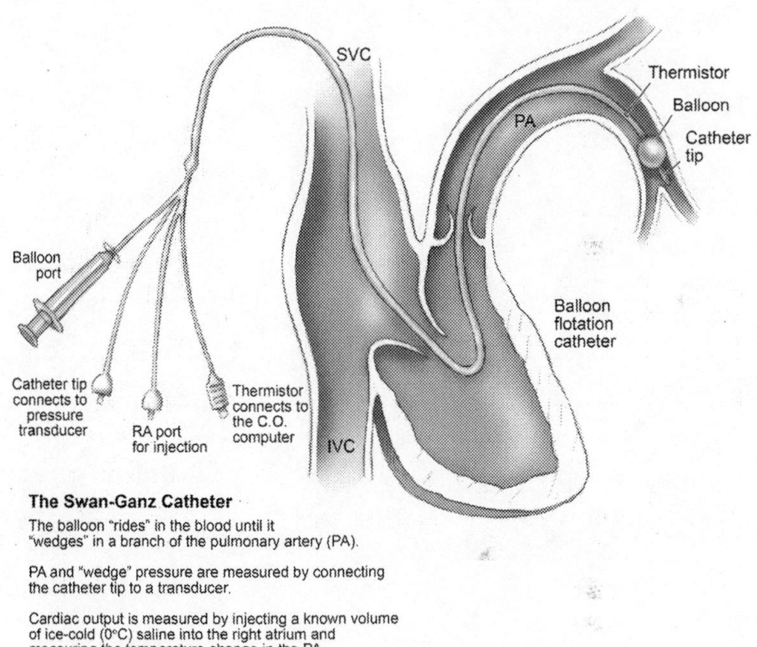


**The Swan-Ganz Catheter**

The balloon "rides" in the blood until it "wedges" in a branch of the pulmonary artery (PA).

PA and "wedge" pressure are measured by connecting the catheter tip to a transducer.

Cardiac output is measured by injecting a known volume of ice-cold (0°C) saline into the right atrium and measuring the temperature change in the PA.

I hope that these figures will help to clarify some of the technical issues in the story.

# Chapter 1

# EDDY BUCZYNSKI [4]

In 1975, as a newly minted cardiologist fresh from Harvard's Peter Bent Brigham Hospital,[5] I started into private solo practice in Worcester, Massachusetts. From my small office on the fourth floor of a relatively new medical office building, I could look directly across busy Belmont Street, Route 9, to the Memorial Hospital emergency department. As part of my staff appointment at Memorial, I was the director of the new coronary care unit. If I walked briskly, and traffic was not too heavy, I could make the journey from the office to the emergency room in roughly two and a half minutes, depending on the traffic coming down the hill. I consulted on a lot of patients with new onset or rapidly worsening (*acute*) heart failure for the ER team.

Heart failure (*HF*) is not a specific disease. HF is what medical doctors call a *syndrome*, a collection of various patient complaints (*symptoms*), evidence on physical examination (*signs*), and *laboratory abnormalities* (on blood tests, electrocardiogram, and imaging studies) that cluster together. The HF syndrome includes a broad spectrum of clinical problems that plague individuals who have impaired heart function. The heart is a pump; its mechanical functions are limited to filling and emptying. Nonetheless, a long list of specific diseases can cause impairment of either its filling, its emptying, or both. If and when the consequences of that reduced cardiac function come to dominate a patient's life, then he or she has the heart failure syndrome.

4 Names have been changed for confidentiality.

5 Now, Brigham and Women's Hospital.

The 1970s and 1980s saw great progress in heart failure research. Laboratory and then clinical data established compelling evidence that the same physiological responses that are activated to retain salt and water under a variety of circumstances like dehydration, gastrointestinal fluid loss (vomiting or diarrhea), or bleeding are also highly activated in heart failure. Collectively, those responses are known as the renin-angiotensin-aldosterone system (*RAAS*). This occurs because the heart is not adequately filled or emptied, and the systemic arterial circulation is underfilled. The relative underfilling that occurs with heart failure stimulates the same RAAS responses that depletion of the circulating blood volume from dehydration or bleeding does. Think of it this way: it's as if you set your home thermostat on the living room wall to a comfortable seventy degrees. Then, when the house temperature was steady at seventy degrees, you take the thermostat off the wall and put it in the refrigerator (without cutting the wires!). Now the relocated thermostat gets a totally inappropriate input because it's cold in the fridge and will fire up the furnace to make the house hotter and hotter. The problem is identical with heart failure; the RAAS is inappropriately activated, like the thermostat in the refrigerator. Patients crave salt and water, and they retain both avidly, like the house getting hotter and hotter. Eventually, the patients become congested. They can't breathe comfortably because of congested lungs, and they gain weight from fluid retention in swollen (*edematous*) legs and ankles. In clinical trials, drugs that blocking the RAAS response (cutting the thermostat wires) produced important improvements in survival for long-term (*chronic*) heart failure patients.

On the other hand, acute heart failure remained a serious problem. A heart failure patient's clinical history is usually a variation on a theme, but every heart failure patient has his or her own individual story with memorable unique details. There is some underlying heart disease—high blood pressure or a previous heart attack. There's a week or two of feeling thirsty and not sleeping well, often in conjunction with some unusual personal stress—maybe a death in the family or a divorce. At first, the patient can't lie flat. He props himself up on two or three extra pillows, or he takes to sleeping in a chair or trying to catnap while sitting at the kitchen table. Then comes a particularly difficult night, the onset of acute heart failure. The patient wakes up short of breath, coughing. Some make it through the night and come to the emergency room in the morning. Some struggle on through the morning and try to take a nap after lunch only to realize that the symptoms are still there. Others don't come in until they get frightened when their sputum turns frothy and pink.

Eddy Buczynski's story was typical; it was also deeply personal for me. Eddy was the all-round manager of the building where I had my new office. He was a wonderful guy. Early on winter mornings, wrapped in his knit scarf, he made sure the plowing service cleared the parking lot, and then he opened a can of cat food for the neighborhood stray and let him chow down in the warm stairwell. He gave the older patients, particularly the ladies, a hand up the three short steps into the building. In the summer, he picked up any refuse that blew over from the convenience store across the street.

"Hi, Doc. You want the place to look professional, right?" he would say, smiling.

Eddy was five foot nine and about 225 pounds, balding, in his early sixties, with adult-onset diabetes and bad high blood pressure. After I had been there a few months, he showed up as a patient. His incredible affection for stuffed cabbage and kielbasa challenged his motivation, but we worked at keeping his problems under control.

Then, the following winter, Eddy's wife got sick. Very sick. After several years in remission, her breast cancer had returned with a vengeance. She was clearly going to die. Eddy paid no attention to himself or to taking his own medications; his life was limited to taking care of the building and taking care of his wife. After a few weeks of this, on a Monday morning, he stopped me on my way into the office, saying, "Doc, I gotta see ya today!"

We rode up together on the elevator. His shoes were untied, and he was sweaty. Sitting on the exam table, his blood pressure was high again, and he was short of breath. When he took a deep breath,

I could hear crackles and wheezes at the bases of both lungs. His jugular veins were distended, and his ankles were swollen.

The acute heart failure patient literally starts to drown in his or her own bodily fluids. The heart's left ventricle can no longer support the circulation without an excessively high filling pressure. Because of the arrangement of the plumbing in normal adult human circulation, the back pressure from the failing ventricle is reflected as rising pressures in the lungs (the "pulmonary capillary wedge pressure"). Forced by the rising lung pressures, fluid seeps out of the vascular bed and into the air spaces (*alveoli*) of the lung. In Eddy's case, his diabetes was probably associated with asymptomatic coronary artery disease, and he had high blood pressure (*hypertension*) as well. Those were his underlying cardiovascular diseases. His shortness of breath and ankle swelling announced that he had developed overt heart failure.

"Eddy," I said, "you've got to come to the hospital. You've got heart failure."

"Doc, I can't. I just can't."

After about ten minutes of verbal head-banging, we agreed that Eddy would spend a few hours in the emergency room, where he got a couple of doses of intravenous Lasix and I renewed the prescriptions for his blood pressure meds.

In 1966, the US Food and Drug Administration (FDA) approval of furosemide (branded as Lasix), a potent, rapidly acting injectable diuretic and vasoactive drug,[6] revolutionized the management of acute heart failure. Before Lasix, in an attempt to reduce pulmonary wedge pressure, heart failure patients were treated with morphine, mercurial diuretics, rotating tourniquets, and occasionally phlebotomy,[7] and some also required sedation and artificial ventilation. Some died. The medical literature of the time did not include large studies, so accurately determining the mortality rate from those days is impossible. After 1966, heart failure could be managed acutely

---

6 By 1973, the important short-term effects of furosemide on the blood vessels were well recognized as clinically important (Dikshit 1973).

7 Basically, bloodletting. In the 1960s, we would draw the blood into a transfusion bag in case the patient needed it back.

with oxygen, small doses of morphine, and intravenous doses of Lasix for diuresis. Diuretics act on the kidney to increase salt and water excretion. After the introduction of Lasix, 95 percent of heart failure patients survived the acute episode. Still, most of them were uncomfortable and short of breath for at least a day or two, and roughly half of them went home with lingering symptoms.

Eddy outlived his wife by about a year and a half; he spent his last six months on disability.

I maintained a private consulting practice in my Worcester office while holding a clinical faculty appointment at the University of Massachusetts Medical School for thirteen years. I cared for hundreds of patients with their own unique presentations of acute heart failure, their own variations of Eddy's story. In 1988, I left private practice to return to a full-time academic position. By the mid-1990s, I was a midcareer academic cardiologist on the faculty of the University of Florida, and Lasix was a thirty-year-old drug. Nothing had changed for patients with acute heart failure.

When I had the opportunity to participate in pharmaceutical clinical research that would help develop a new drug for acute heart failure called nesiritide (eventually brand named *Natrecor*), I was more than ready.

## Chapter 2

# FROM THE DOG LAB TO MOLECULAR STRUCTURE

The chain of scientific discovery leading to nesiritide had begun decades before I reached Florida. The best place to start is in the mid-1950s at a laboratory on the Wright-Patterson Air Force Base in Dayton, Ohio. A veterinary surgeon, Captain J. L. Reeves, had joined Drs. J. P. Henry and J. W. Pearce to lend his skills to their research efforts. In the lab, gowned and gloved, Reeves neatly opened the chest of an anaesthetized dog that was lying in a surgical cradle with a breathing tube and a bladder catheter in place. Without interrupting the flow of blood through the lungs, he gradually expanded a balloon in the holding chamber for the right heart (the right atrium[8]). The expanding balloon simulated the effect of blood volume expansion on the heart by mechanically stretching the atrium without actually giving the dog additional blood or fluid. The catheter in the dog's bladder soon started to fill with urine.

The experiment showed that stretching the right-sided holding chamber of a dog's heart with a balloon made the dog pass more urine (Henry and Pearce 1956). The inescapable conclusion was that by some as-yet-unknown mechanism, the heart had sent a signal to the kidneys. How?

---

8 *Atrium* is Latin for "waiting room." The atria of the heart are where blood "waits" to enter the ventricles for pumping to the lungs or to the body.

As with most stories, this starting point is arbitrary. Henry and Pearce were looking for something with their experiments. What was it, and why were they searching?

By 1950, many of the questions about the structure of the body had been answered. Anatomy, except for the description of subcellular structures like the tiny energy-generating structures inside the cell called mitochondria, had largely solidified into a fixed body of knowledge. In biochemistry, the basic processes were also largely understood. Although for biochemistry, new breakthroughs—particularly the dramatic discovery of the structure of DNA—were still ahead. For researchers in physiology, the biomedical discipline that deals with the engineering questions of "how does it work?" the body remained full of terra incognita to explore. One of the critical physiologic questions that had started to yield preliminary answers was, "How do animals, particularly dry-land animals like us, regulate the salt and water balance of our fluids to maintain the chemistry of our interior environment constant at levels fairly similar to seawater?"

Some hints to the answer had come from the observation that when a disease like tuberculosis destroys the adrenal glands (the small, fleshy organs that sit just above the kidneys), patients easily lose excessive amounts of salt and water. Their inability to retain salt and water leaves them vulnerable to perilously low blood pressure following minor insults, like gastroenteritis, that healthy individuals tolerate relatively well. Clinically, the condition is called is called Addisonian crisis.[9] Early on, the adrenal glands were recognized as an essential part of the critical regulatory pathway that stimulates salt and water retention.

As the details were worked out through multiple experiments, the regulatory pathway became well known as the RAAS, which is so important in heart failure. It acts as a powerful accelerator to increase heart rate, drive salt and water retention, and raise blood pressure.

---

[9] During the 1960 campaign, Kennedy's opponents said he had Addison's disease. His physicians released a cleverly worded statement saying that he did not have Addison's disease caused by tuberculosis, and the matter was dropped (Maugh 2009).

But physiologists had come to understand that there must be a counterregulatory mechanism in order to balance the RAAS. There had to be a brake that could cause reduction in blood pressure and encourage salt and water excretion. Physiology regulates processes with both positive and negative controls, just as the driver of a car has to use both the accelerator and the brake to control the vehicle's speed.

That day in 1950, Henry, Pearce, and Reeves knew at least the approximate location of the accelerator; they wanted to learn more about the brakes.[10] The results of their experiment showed that the heart could signal changes in its volume to the kidney; they had the first big link in what would grow to become a chain of evidence.

Fast-forwarding to 1983, and with a change in venue from Dayton, Ohio, to Canada, the next critical link would come from Adolfo de Bold's physiology laboratory at Queen's University in Kingston, Ontario. De Bold, a brilliant Argentine émigré, had developed a technique to study the function of an isolated perfused rat kidney.[11] As he described it, "The kidney rested in a water-jacketed kidney holder and the perfusate was kept at 37.5°C[12] and oxygenated with 95% O2 - 5% C02" (Baines, de Bold, and Sonnenberg 1983). In other words, this preparation totally isolated the kidney, keeping it alive but independent from the rest of the body.

Based on the experiments in Dayton and additional later work, de Bold had worked out a technique to homogenize rat heart atrial tissue. He had then partially purified the homogenized material.

---

[10] Why was the air force involved? After WWII, aeronautical engineers developed practical high-speed jet aircraft. In aviation, human factors had entered the process as practical limitations. Maintaining the pilot's consciousness while maneuvering high-speed aircraft demanded new techniques for maintaining blood pressure and flow to the brain. Aerospace medicine and physiology have advanced hand in hand in understanding many regulatory systems.

[11] Since the kidney had been removed from the body, there was no blood flow to supply nutrients and oxygen. Instead, de Bold substituted an artificial fluid of known composition called a "perfusate" to keep the kidney working.

[12] Body temperature.

When he injected the partially purified substance—whatever it was—intravenously into rats, it consistently caused changes in the composition of their urine.

Did the unidentified substance in the homogenized tissue act directly on the kidney? Did something in his tissue preparation carry the message that Henry and Pearce had shown that the heart sent to the kidney? De Bold wanted to know.

With his isolated perfused kidney, he could duplicate the effects of the expanded balloon with the atrial tissue extract. He was closing in on a signal that counteracted salt and water retention. He injected increasing doses of the partially purified material into the fluid perfusing the kidney. As he did, the isolated kidney produced urine with increasing amounts of salt and water in response.

In his words, "In conclusion it appears that AE [atrial extract] is a potent natriuretic, chloruretic, phosphaturic, and diuretic in isolated perfused kidneys. Its action does not involve alterations in catecholamine excretion" (Baines, de Bold, and Sonnenberg 1983).

So far, the science showed that the atrial muscle cells of the heart contained a chemical messenger. When the cells of the atria were stretched by a balloon or by expansion of the circulating blood volume, they released that messenger into the bloodstream. The messenger, whatever it was, caused blood vessels to dilate. However, de Bold, with his isolated kidney preparation, had proved that independent of the blood vessel effects, the messenger substance also directly increased salt and water excretion by the isolated kidney. In physiologic terms, the messenger was a salt excretion (*natriuretic*) hormone. The outlines of the regulatory system were growing much clearer.

Just two years later, de Bold (de Bold 1985) described purified atrial natriuretic factor, or A-type natriuretic peptide (ANP) as we now call it.

The final scene in this process of continuing discovery occurred around the world from Dayton, in Japan in 1990. Yoshikazu Kambayashi and his associates worked in a cooperative research venture involving the venerable Japanese pharmaceutical company

Shionogi and Kyoto University. Together, they reported the isolation of B-type natriuretic peptide (BNP) from human heart tissue and the determination of the sequence of amino acids in the molecule. In their words, "In the present study, we have isolated human BNP from the human atrium and demonstrated that human BNP is a 32-amino-acid peptide, which corresponds to the C-terminal sequence (77-108) of the human BNP precursor deduced from the cDNA sequence" (Kambayashi et al. 1990). They had finalized the chain of evidence. They had found the cardiac messenger, determined its chemical structure, and located the genetic information that coded its production.

There are, as it turns out, several different natriuretic hormones. They are all made up of chains of small molecules called amino acids linked by specific chemical bonds, called *peptide* chains. The amino acid sequences in the peptide chains differ slightly from species to species, but the system of salt and water regulation is far too physiologically important to permit much variability. Technically speaking, the molecular structure is *highly conserved* across species. The most important natriuretic hormones in human cardiac physiology are type A, or atrial natriuretic peptide (ANP), and type B (BNP).[13] The specific sequence of amino acids for each natriuretic peptide is coded by our DNA.

---

[13] Type B was first isolated from porcine brain, hence the early use of the term "brain natriuretic peptide." The heart was subsequently identified as the primary source for type B, and "brain natriuretic peptide" became an anachronism.

## Chapter 3

# BIOTECHNOLOGY

While the cardiorenal physiology of nesiritide was being unraveled slowly and rather quietly, the world of biochemistry exploded. In 1953, Watson and Crick reported the complementary base structure of DNA (Watson 1953).

Over the next twenty years in laboratories around the world, researchers labored to understand the details of exactly how the DNA code was read and how it dictated the structure of the peptides and proteins[14] that are the key structural and chemical operating units of the body. Then, in 1973, two West Coast scientists—Herbert Boyer from University of California–San Francisco and Stanley Cohen from Stanford—developed a technique by which they were able to isolate specific frog DNA, insert it and clone it in bacteria, and show that the bacteria could use it as the molecular instructions to manufacture frog protein. They had invented *recombinant DNA technology* (Hughes 2011).

With the twenty-first century now well on its way, it's hard to remember what these discoveries first meant in context. Because DNA codes for specific amino acids, if you know the sequence of purine and pyrimidine bases in a section of DNA, then you can use the code to predict the sequence of amino acids in the peptide or

---

[14] Without getting deeply into technicalities, the difference between a peptide and a protein is roughly analogous to the difference between a large piece of gravel and a boulder. It has much more to do with size than composition.

protein it codes for. Or you can go the other way; if you know the amino acid sequence of a peptide hormone, you can predict the sequence of bases in the DNA that codes for it.

Robert Swanson, a young San Francisco venture capitalist who had an MIT undergraduate degree in chemistry, saw the enormous business potential of the new technology. Along with Boyer, he founded a new company; they named it Genentech (Hughes 2011).

A few years later, William Bowes, a Stanford alumnus, and Winston Salser of UCLA formed the nucleus of another biotechnology company and incorporated it as Applied Molecular Genetics (Amgen) in 1980 (Binder and Bashe 2008). The company was headquartered in Thousand Oaks, California. They recruited George Rathmann, an Abbott executive with extensive Big Pharma experience, as the new company's chief executive.

As Genentech and Amgen developed their businesses, they showed that they could use recombinant DNA technology to manufacture human peptide hormones in quantity. The cost and complexity of the drug business, however, pushed both of the start-ups to partner with large, established companies to get a foothold in the industry. For its first commercial product, Genentech partnered with Lilly in 1978 to develop recombinant human insulin as Humulin. FDA approval of Genentech's second product, human growth hormone, developed with Kabi, came in 1985. Amgen partnered with Kirin and J&J (more about that relationship later) to develop erythropoietin (EPOGEN, PROCRIT), which was FDA approved in 1989.

The trail that Genentech and Amgen had blazed for small entrepreneurial biotechnology companies to succeed in the pharmaceutical industry consisted of four steps. The first three steps are increasingly expensive. The fourth, following a huge *if*, is the payoff:

1. Identify a peptide hormone that plays a key role in an important human disease and establish its structure.
2. Clone or synthesize the DNA sequence of the hormone in order to biomanufacture pharmaceutical amounts.

3. Take the resulting recombinant hormone through the process of clinical drug development as a potential treatment for the disease.
4. *If* step 3 is successful, find a partner to market the hormone as a new drug.

By 1990, the concept that naturally occurring human peptide hormones could be repurposed as therapeutic agents, as drugs, had been well validated. Genentech and Amgen had turned scientists into wealthy businessmen. Certainly, the idea of testing the naturally occurring natriuretic peptide hormone B-type natriuretic peptide (BNP) as a therapeutic agent to help restore salt and water balance in heart failure patients with acutely decompensated heart failure was a mainstream business concept in the San Francisco Bay Area.

Did nesiritide have a future as a new heart failure drug? Was it, perhaps, a useful part of the braking system for the heart failure syndrome? A small company called California Biotechnology thought it might be.

## Chapter 4

# EARLY CLINICAL TRIALS

Given the proximity of Stanford, UCSF, and the biotechnology industry, finding new medicines was a pervasive multimillion-dollar dream in the Bay Area. The ultimate dream, of course, was to develop a blockbuster—a drug that generates a billion dollars in sales in one year. It had happened for Genentech, and it had happened for Amgen. It was possible.

California Biotechnology had become Scios for a short time in the early '90s and then it renamed itself Scios Nova Inc. after merging with Nova Pharmaceuticals in 1992. It would soon drop the *Nova* to become simply Scios Inc. The name was intended to vaguely suggest a sort of Greek or Latin reference to "knowledge." No matter what the name was, the company managed to keep the lights on with venture funding and contract research while it pursued the Bay Area dream.

For Scios, pursuing the dream with nesiritide meant that the company would have to begin that third step on the path toward becoming a pharmaceutical company, the big-money and big-risk step: the rigorous process of development and clinical research to find out if nesiritide could be developed into an approvable drug.

Unfortunately, Scios was coming late to the party. Biotechnology had become an industry by commercializing recombinant human peptide hormones as drugs. The companies that had early successes had already picked the low-hanging fruit; by the 1990s, there were not a lot more common diseases that could promise big returns for a peptide hormone treatment. After all, the peptide hormones had to be given intravenously, and they had to meet the FDA requirements for both safety and efficacy. Scios did not have a plan B, a backup compound in some other area. For the company, nesiritide was it—all or nothing.

Scios had enough chemically synthesized nesiritide to carry out the animal toxicology and safety pharmacology studies that must be done with every compound as it starts down the pathway of drug development. As with other recombinant peptide hormones, by inserting a short section of DNA that coded for human BNP into bacteria (*E. coli*), the bacteria turned into microscopic BNP factories. This is the marvel of recombinant technology, and with recombinant manufacturing, nesiritide became a true biotechnology product. It was now recombinant human B-type natriuretic peptide, or rh-BNP.

* * *

Before I was involved, the chemically synthesized BNP Scios made had already been officially named *nesiritide*. In addition, the preclinical and phase 1 drug development work had been completed with chemically synthesized nesiritide. This included giving it in large doses to animals to look for toxicity and then the phase 1 "first in human" studies in healthy normal subjects to establish initial dose ranges. In addition, the Scios clinical development team had worked with academic heart failure specialists to do some very limited phase 2 work in patients with heart failure, first with single doses and then with short infusions lasting four to six hours. These early nesiritide investigators included heart failure specialists Milton Packer and Stu Katz at Columbia University in New York City, and Robert Hobbs and his associates from the Cleveland Clinic.

When recombinant nesiritide became available in quantity, the next step was a larger phase 2b trial. In drug development, phase 2 is a critical, make-or-break stage. It's the first time that patients, rather than healthy research subjects, actually get a new drug across a wide range of doses and under conditions where the development team can look at standard measures of efficacy and safety. For Scios, this meant going to a dozen or so medical centers around the country to actually find out whether giving nesiritide as an intravenous infusion to heart failure patients would work.

Of course, to do this, the research team first had to agree on what effect (or *parameter*, to be more formal) they would use to indicate that nesiritide was working. The parameter (or parameters) they chose had to be acceptable to the FDA as an indicator of the severity of heart failure; in addition, whatever they measured had to change within a few hours in response to nesiritide, and the change had to be objectively measured.

In the mid-1990s, this set of requirements limited the choices to just one approach, using a special Swan-Ganz right heart catheter to directly measure the pressures required to push blood through the lungs into the main pumping chambers of the heart and to measure the actual volume of blood that the heart was circulating, the cardiac output.

The relationship between the filling pressure of the left heart and the volume of blood it ejects is a critical measure of the performance of the healthy and the failing heart. In the early '70s, two researchers, cardiologist Jeremy Swan and physiologist Willie Ganz, developed a thin, flexible catheter with a balloon on the tip that could be used to monitor filling pressures and cardiac output at the bedside of an acutely ill patient (Forrester 2015). The technology was quickly adopted and led to major advances in the management of patients who had suffered heart attacks (*acute myocardial infarction*). By the early '90s, the Swan-Ganz catheter was widely used for tailoring drug treatment for patients with heart failure, particularly in hospitals with heart transplant programs.

For most of 1997, 16 heart failure centers had joined in the effort to recruit 103 patients for a study known officially as Scios 704.311, or 311 for short. By 1998, the work was done. The data that were collected in at Scios's offices in Sunnyvale, California, included scores of pages of computer printouts with the critical information about the people who had agreed to participate in the study. There were also yards and yards of red-lined medical graph paper with the wave forms of the Swan-Ganz pressure records.

Now someone comfortable with interpreting the pressure waves recorded on those yards of paper had to review the data for a quality check and then put all the results on the computer printouts together into a manuscript that would be suitable for scientific review and publication. That someone was me.

As luck would have it, I had met the tall, patrician, and genial Robert Hobbs in Chicago at an international heart failure and transplant meeting. Over a glass or two of wine, he told me about "an interesting new compound" that the heart failure group at the Cleveland Clinic had used in a small study. When I confirmed my interest, he introduced me to the Scios drug development team. By that time, I was working at the University of Florida College of Medicine, where Dick Conti and Carl Pepine had developed an experienced and highly competent clinical research program. With the help of two dedicated research nurses, Kay Price and Dana Leach, the UF program recruited more patients into the 311 study than any other single center. As was the custom at the time, the data review and writing would be my job.

The Scios campus in Sunnyvale sat about thirty miles south of the San Francisco airport, just off US 101. The low, redbrick buildings looked tidy but unimposing. The cafeteria buzzed with the conversations of healthy, tanned people wearing open collars, jeans, and khakis. The window opened on a new world. I was an East Coast alien, pale, dressed in a tie and blazer; they were the natives, bright, intense, and enthusiastic.

As exciting as it was, I wasn't there to enjoy the Bay Area. I was there to put the data together into a major clinical research paper. Without Darlene Horton, it would have been impossible.

Darlene had trained as a pediatric cardiologist. She was petite, and she was always alert; sometimes she seemed like a small bird of prey that was ready to hunt. She had been with Scios through almost all the development process. Not only did Darlene know everything about nesiritide, she had near-photographic memory for the details of every previous study: who did it, what the data showed, where and when it was published, and how it added to the big picture.

We spent three intense days working on a draft manuscript. For a busy cardiologist, it was real "California dreaming." There were no patient-related calls, no emergencies, just reviewing the data and writing.

Actually, there was one emergency. One of the scientists in the building, a smoker, developed some rather suspicious chest pain. Elliott Grossbard, who had trained as a hematologist and had come to Scios from Genentech, brought the poor fellow over to see me. All I could do was to second the recommendation that he should get to an emergency room right away. In fact, he did have an acute coronary event.

Clinical research manuscripts always follow a conventional format: introduction, methods, results, discussion, and conclusion. The investigators' brochure and the research protocol[15] already included much of the required text for a research paper. With just the one interruption and ready access to Darlene, who could reliably fill in key details, I had the initial draft completed and ready to send to the collaborating investigators for review by the afternoon of the third day. Early that evening, I joined Darlene and her partner to share a bottle of well-chilled California chardonnay. As we enjoyed the Bay Area breeze on the balcony of their second-floor apartment, the path forward seemed clear.

As the pieces of the puzzle came together, the phase 2b data from the study showed that, at baseline, the patients had severe advanced heart failure as the protocol required. They looked much like Eddy had on the morning he first went to the hospital. When they received intravenous (IV) infusions of nesiritide, they had experienced rapid reductions in the pressures in the lung (*pulmonary*) circulation, and their forward cardiac output increased. Importantly, the patients maintained their improvement throughout the twenty-four hours of required exposure to the drug. A fairly large percentage of patients who received a high dose of nesiritide experienced what we considered excessive drops in their systemic (arterial) blood

[15] These documents must be completed before the study begins.

pressure;[16] however, we saw a sweet spot in the dosing that produced significant improvements in pulmonary pressures and cardiac outputs without too much systemic blood pressure reduction. The study results looked good.

The next step would be to discuss the data with the FDA and agree on an acceptable design for a larger phase 3 study, with more centers and more patients. Given the unmet medical needs of patients with heart failure, if the phase 3 trial was as successful as the study we had just finished, nesiritide had a good chance of gaining FDA approval for treatment of acute heart failure.

That was the last time the way forward was going to seem clear.

---

16 This is the blood pressure that is usually measured with a cuff around your arm.

## Chapter 5

# THE FIRST ROUGH TIMES

A few months after we completed the manuscript for the 311 twenty-four-hour infusion paper, Darlene called. She invited me to participate in a face-to-face meeting with the cardiorenal team at the FDA. The purpose of the meeting at the end of phase 2 was to obtain FDA guidance on planning the definitive phase 3 study. It was critical that the agency agreed beforehand that, if the phase 3 study were successful, it would provide the evidence required for the FDA to approve nesiritide for marketing.

In these meetings, the company developing a new drug is referred to as the sponsor. She explained that the sponsor's team would include her as the head of clinical development for Scios, me because of my role as first author of the 311 paper, and Wilson Colucci, who was to be the principal investigator for the phase 3 study.

Wilson, known as Bill to most of the cardiology world, had completed his cardiology training at the old Peter Bent Brigham Hospital[17] just a few years after I did. He has spent his professional life in Boston, rising to become professor and chief of cardiology at Boston University. For the decades that I have known him, Bill has always fulfilled the image of a prominent Boston academic physician: fit, with neatly trimmed short hair, a tweed sport coat, and round, wire-rimmed glasses. By the late 1990s, Bill had gained

---

[17] The Peter Bent Brigham Hospital opened in 1913 and merged into the current Brigham and Women's Hospital in 1980.

wide recognition as an academic cardiologist with special expertise in heart failure.

In contrast to Bill and Darlene, I had no experience with regulatory issues, no real idea how the FDA worked, and limited credibility as a researcher. Participating in a meeting with the FDA as a member of the sponsor's team presented a unique opportunity to learn more about drug development. I jumped at the chance.

At the time, the FDA sat near the NIH, across the Rockville Pike from the National Naval Medical Center (NNMC) in Bethesda, Maryland. We arrived on a cold, rainy, late-winter day. I had done a couple of two-week US Navy Reserve tours as a medical officer at NNMC–Bethesda. The FDA had the same familiar drab federal interior decorating. We had a monochromatic meeting room painted in a shade that does not occur in the natural world. The room was furnished with folding chairs and a battered table. One of the senior FDA people sat down. He looked at the assembled group and started the meeting by bluntly growling, "So this is about a new drug for acute heart failure. You treat that with rotating tourniquets,[18] don't you?"

In the stunned silence of the moment, the right lens of Bill's glasses fell out and rolled across the table. No one knew what to say. Thirty years before this meeting, Lasix had made rotating tourniquets passé.

The question was simple and somewhat ingenuous, but it had profound implications. The FDA wanted to start with the basics. Bill recovered his lens, and with it, his composure. Then we spent the next hour discussing how treatments for acute heart failure actually worked and what a phase 3 acute heart failure study should look like. We agreed on the broad outlines of what would become two studies, the nesiritide safety and efficacy trials (officially, Scios 704.325 and 704.326, or 325-326 for short). The trials would eventually include

---

[18] "Rotating tourniquets" referred to the practice of putting a tourniquet with enough pressure to occlude venous return on three of a patient's four limbs. A 1974 publication noted, "This form of treatment has been used for many years," and cited references from 1928 and 1942 (Habak et al. 1974).

66 centers in the United States with a total of 432 patients, and the combined results would provide the pivotal data for the sponsor's January 1999 application for FDA marketing approval.

There were four mutually agreed-upon goals for the combined studies.

1. Using Swan-Ganz catheterization, the studies would provide more data on cardiac filling pressure (*pulmonary capillary wedge pressure*[19]) and cardiac output (*hemodynamic*) changes in response to nesiritide in heart failure patients with what we believed would be optimal doses of nesiritide.
2. The studies would then go on to show that Swan-Ganz catheter monitoring was not required in order to administer the drug safely.
3. Third, we would gather data to show whether patients who received IV nesiritide along with their usual care would have more rapid relief of their symptoms than patients getting usual care and an IV placebo.
4. The studies had to be large enough to substantially increase the amount of safety data related to the use of nesiritide in acute heart failure patients.

Two of the four goals were both particularly difficult and critical for success. One was choosing a dose of the drug that would be reliably effective in treating the majority of patients without the investigator having access to measurements of filling pressures and cardiac output.

Dosing is always a challenge in designing phase 2 studies. It is a classic Goldilocks problem: not too high, not too low. The sponsor

---

[19] The "pulmonary capillary wedge pressure" is an easily measured and reliable reflection of the filling pressure for the left ventricle, the main pumping chamber of the heart. Dr. Lewis Dexter, who described the measurement of pulmonary capillary wedge pressure in 1947 (Dexter et al. 1947) had been a mentor to Bill and me during our training at the Peter Bent Brigham.

wants to get it just right. Sometimes it's possible to pick a single dose, but often, as with nesiritide, there just isn't enough data, and the sponsor has to include several doses. On top of that, we and the FDA knew that nesiritide could never achieve widespread use if every patient who received it had to have a Swan-Ganz right heart catheter placed in order to adjust the drug-infusion rate. The time and expense involved in placing a catheter, the need for specialized care in an intensive care unit bed, and the rare but real complications of Swan-Ganz catheterization combined to make it an unacceptable strategy for everyday use. We would have to decide on an infusion rate that would work for almost every patient and then specify modest subsequent dose adjustments that would be appropriate for those patients who either had blood pressure drops or didn't improve.[20]

The second critical problem was to convincingly demonstrate that, in addition to improving the filling pressures and cardiac output, nesiritide actually helped patients feel better. The rules of the game at the FDA require a sponsor to show that a new drug either improves symptoms or increases survival. That's how the agency defines *efficacy*. No one thought that a short course of nesiritide would increase long-term survival, and besides, testing long-term survival would require an unacceptably long study. With these constraints, the symptomatic outcome had to be the primary efficacy measure, and the specific symptom of interest was the shortness of breath that accompanies heart failure, referred to as *dyspnea*. The sponsor's team (consisting of Darlene, Bill, and me) was naively confident that if we had a drug that reduced the pressures in the lung circulation and increased the cardiac output, the patients would certainly soon feel better. After all, we had seen it happen at the bedside time after time, and I had the data from the 311 study as well. The FDA regulators simply said, "Okay, prove it."

---

[20] Correct dosing is a major challenge in new drug development; the challenges of developing computer modeling for drug dosing have attracted at least one former rocket scientist (my former colleague Ihab Girgis) to pharmacology.

* * *

Darlene and Bill worked together to design the trials, write the protocols, and put together a strong group of investigators and sites. Bill was an effective leader for the trials. He leveraged the widespread interest in nesiritide among HF specialists. The prospect of testing a novel drug for acute heart failure interested many key people in academic centers. Even more important, the prospect that the combined trials would produce data that might win rapid FDA approval led Dick Brewer, the Scios president, to bet the farm and spend the limited funds that were available freely to get it done.

The upshot of this combined corporate and academic enthusiasm was that the 325-326 studies were completed in the nearly ten months from early October 1996 to late July 1997. This was extraordinarily speedy for a phase 3 trial; at the usual recruiting rates for similar trials, the studies would have required at least two years. In addition, the expense of the endeavor severely stretched the limited resources of the company. Clinical trials cost a lot of money. This will be an issue later in the story; for now, it's just a fact.

Bill Colucci, Darlene, and the ten leading site investigators would later publish the full report of the combined 325-326 trials in the *New England Journal of Medicine* (Colucci et al. 2000). Among the 127 patients who had Swan-Ganz catheters placed, 85 received nesiritide and 42 received placebo. A moderate dose of nesiritide (0.015 μgm/min) caused the pulmonary capillary wedge pressure (PCWP) to decrease by about 20 percent, and the higher dose of nesiritide (0.030 μgm/min) decreased it by almost 30 percent as compared to the baseline pressures before treatment. The patients receiving placebo had a very slight increase in their pressures.

The publication also described the outcomes for 305 patients who were randomized in the comparative trial that was designed to compare symptom improvement with nesiritide versus "standard therapy." These patients did not have to undergo Swan-Ganz catheter placements. The 102 "standard care" patients received a variety of potent intravenous drugs, including dobutamine (57 percent), milrinone (19 percent), and nitroglycerin (18 percent). The 203 nesiritide patients were divided into 103 at the moderate dose and 100 at the high dose. At the end of the day, there were no significant differences in any of the end point efficacy measures: dyspnea, fatigue, or global clinical status, between any of the treatment groups.

Putting these results into context, the nesiritide infusions clearly had an impressive hemodynamic benefit as compared to a placebo infusion. On the other hand, when nesiritide was compared to other widely available generic drugs that the study investigators used for standard care, its effect on how patients felt was essentially the same. However, as the authors pointed out, nesiritide was a lot easier for the investigators to use. Two of the standard heart failure drugs used for comparison, dobutamine and milrinone, may cause rapid or irregular heart rates and other arrhythmias. Nitroglycerin, the other commonly used comparator treatment, causes headache for many patients, and most patients also rapidly develop tolerance to the drug, requiring the nursing staff to make constant dose increases.

The clinical results might have been similar in the study, but for doctors, nurses, and patients, nesiritide was a much easier, simpler drug to manage. As the investigators stated in their paper, "The salutary clinical and hemodynamic profile of nesiritide and the relative absence of adverse effects associated with it circumvent several [of these] limitations [of standard drugs]. We therefore suggest that nesiritide would be a valuable addition to the initial treatment of patients admitted to the hospital for decompensated congestive heart failure." That was the essence of the package that would go to the FDA early in 1999.

* * *

Analysis of the 325-326 data and preparation of the FDA application occupied the Scios research team for much of 1998. Meanwhile, on the business side, Dick Brewer established his leadership at Scios. His arrival in 1998 had energized the company's management. He was a high-energy, larger-than-life figure with experience and contacts throughout the pharmaceutical industry. Dick had thrived on meeting the challenges of a growing biotechnology endeavor. In just over a decade with Genentech, he had risen to senior VP of sales and marketing and then to senior VP for Europe and Canada. At Scios, he planned to follow the pathway to success that had become standard for fledgling biotechnology companies. While the development team got the data ready for the FDA, Brewer quietly started putting in place the framework for a clinical partnership. He worked out a cooperative deal with the German giant Bayer that would solidify the company finances and move nesiritide into the marketplace. The Bayer deal was contingent on the first-round FDA approval of nesiritide as a commercially branded drug, to be known as Natrecor.

* * *

The FDA Cardiovascular and Renal Drugs Advisory Committee[21] met on Friday, January 29, 1999, to consider the approval of intravenous Natrecor for the treatment of acute decompensated heart failure (US FDA 1999). The agency had not reviewed a new IV drug for heart failure in the past twelve years. Milton Packer, a gangly and often rumpled academic who sometimes moved as if he and his body had not achieved consensus on their destination, chaired the meeting; two academic heart failure specialists, Marv Konstam and Joann Lindenfeld, were among the committee members. Rob Califf, a cardiologist and an outstanding clinical researcher at Duke, was on the committee but did not attend the meeting.[22] For the sponsor, Darlene Horton led the Scios contingent. A rising young academic cardiologist, Bill Abraham[23] from the University of Cincinnati, who

---

[21] In his book *New Drugs*, Friedhoff says, "It is a shame that more of the public does not get to see advisory committee meetings … the public would have much better understanding of the challenges that face FDA … if these meetings were available" (Friedhoff 2009).

[22] Califf is now (2015) FDA deputy commissioner for Medical Products and Tobacco (US FDA 2015c).

[23] Abraham is now professor and director of the Division of Cardiovascular Medicine at the Ohio State University (Ohio State University Wexner Medical Center 2016).

was the fourth coauthor on the primary study publication, appeared as the sponsor's academic expert.

The conflict-of-interest section of the minutes contained a disclaimer. "Dr. Packer would also like to note that he was involved in the early development of Natrecor, as a consultant, but his participation in the program ended more than two years ago. Dr. Packer's employer, the Columbia University College of Physicians and Surgeons, was involved in a phase II study of Natrecor. Dr. Packer was listed as an investigator on the study but did not participate in the recruitment of patients, or the analysis of the data" (US FDA 1999). Packer had been the third author on the publication that had reported the Columbia group's findings (Marcus et al. 1996). No matter what the disclaimer said, he knew the compound well.

When the meeting started, the Scios team thought the ground rules for the game were clear. As agreed with the FDA in the planning meeting before the study, the reduction of filling pressures measured with the Swan-Ganz catheter was an appropriate primary end point. Changes in cardiac output, blood pressure, and clinical status were also reported as secondary end points. The data showed that nesiritide could be used safely and effectively without a Swan-Ganz catheter, with clinical results similar to standard care. Finally, the total safety experience with Natrecor now included approximately five hundred patients.

But Milton Packer, speaking from the chair, quickly called that agreement into question. "This Committee has not seen an IV drug for the treatment of heart failure in 11 years. So it is not clear how all of these guidances should, in fact, be incorporated. So I think most importantly we need to look at the data, and see what the data, in fact, would indicate to us" (US FDA 1999, 12).

Before the committee had looked at any data, the chair had effectively declared that all bets were off, and the rules of the game were suspended. The planning from the end-of-phase-2 meeting with the FDA, the four agreed-upon objectives of the 325-326 studies that had been so carefully hammered out and built into the research protocols, and two years of clinical research and data analysis were now suddenly in limbo.

Robin Allgren from Scios presented the efficacy data from the 325-326 studies and the dosing recommendations for the sponsor. The efficacy data included the hemodynamic data, and both patients' and physicians' assessments of improvement. She concluded that Natrecor had "the characteristics desirable for an IV vasoactive agent for the short term treatment of CHF [congestive heart failure] with beneficial effects on both hemodynamics and clinical status." The sponsor's recommendation to the committee was a fixed-dose infusion at 0.015 µgm/kg/min without a loading dose (US FDA 1999, 13). In the double-spaced transcript, Robin's presentation wrapped up on page 37. Question-and-answer sessions at FDA advisory committee meetings usually last ninety minutes. The Q and A following her presentation ended on page 163. She and her team had faced roughly two hours of continuous questions.

Some of the questions were technical, but surprisingly, many were related to the details of how the patient-reported changes in dyspnea were ascertained. At one point, justifying the degree of attention to dyspnea, the FDA's Ray Lipicky commented, "This idea that you get the approval of a new treatment because you can demonstrate that you make people feel better, or live longer, or both ... is a pretty fundamental notion. And if wedge pressure is the only thing you are looking at, you are indeed looking at a surrogate" (US FDA 1999, 146).

The term *surrogate outcome* had not been prominent in previous discussions, but it was going to become so now. Clinical research in cardiovascular disease, and particularly in heart failure, often measures the impact of a drug, device, or procedure on survival rates, often called *hard outcomes*. In other chronic disease studies, symptomatic improvements—or *patient-reported outcomes*—are accepted end points if measured rigorously. By stating that pulmonary capillary wedge pressure was a surrogate, Lipicky meant that it was neither a patient-reported (i.e., "feel better") outcome nor was it demonstrably equivalent to the hard outcome of living longer. He had succinctly stated one of the most critical issues in all the controversies that would subsequently develop around Natrecor. If Natrecor treatment did not improve survival after an episode of acute heart failure,[24] then the efficacy measure that would support FDA approval had to be symptomatic improvement. The way that dyspnea would be measured had not been addressed in great detail in the end-of-phase 2 discussions. But as the chair had warned, that was not going to count in the day's deliberations.

Darlene then came forward to present the safety data. From the overall data, she concluded that nesiritide could cause dose-related decreases in systemic (arterial) blood pressure. She also went into detail in discussing the patients who died during the trials and emphasized the complexity of their cardiovascular problems. Then she went on to discuss the impact of nesiritide on kidney function, saying "when we looked at patients [receiving nesiritide] who had … either an increase in creatinine of more than 100% or patients who developed acute renal failure requiring dialysis … there is no difference in the frequency of these events compared to control" (US FDA 1999, 185).

She also pointed out that in the control group for the 326 study in which patients received other "standard" intravenous heart failure drugs, 3 percent of the patients had a cardiac arrest during treatment (all of them were receiving dobutamine); none of the nesiritide patients had an arrest during treatment.

Bill Abraham then gave a summary that outlined the proposed clinical use for nesiritide, saying, "Candidates for treatment with Natrecor are hospitalized patients with decompensated heart failure requiring intravenous vasoactive therapy … volume overloaded, and not in cardiogenic shock."

Another hour or more of questions followed the safety presentation.

After an intense discussion, the committee voted to recommend approval for marketing: five in favor, and three (Konstam, Moye, and

24 Importantly, Scios had never claimed, or even hoped, that short-term treatment with Natrecor would have an impact on long-term (six months or more) survival.

Pina) opposed. Then, after three months of silence, on April 27, the agency announced that it had not accepted the advisory committee's recommendation and would not approve Natrecor. The regulators had decided that they wanted more data, in particular more safety data (*Pink Sheet* 1999).

Given the role and responsibility of the FDA, the leaders at the agency had made a sound decision; later events would bear that out. But at the moment, it seemed as if Scios had scored a touchdown. Then, while the crowd cheered and the cheerleaders danced, the referees had quietly moved the goal line.

With the announcement that Natrecor would not be approved without additional phase 3 data, Bayer withdrew from the planned partnership agreement. Like countless other small companies, Scios had discovered the unpredictability and high costs of clinical drug development. Now the question in front of management was simple and straightforward: "Do we call it a day, or should we raise the money for another trial?"

*Chapter 6*

# BREWER ON THE ROAD

By 1999, I had departed from Florida and had joined the staff at the Cleveland Clinic.

"There's a call from California for you. Can you take it?" Maddy asked.

Maddy Just organized my professional life at the Cleveland Clinic, and she had an uncanny sixth sense for what was important.

"A Mr. Brewer would like to meet with you and Dr. Hobbs[25] next week. Can you make it?"

"I'll be happy to," I answered. "I'll pick it up."

I grew up in the gray, damp of northeastern Ohio, not too far from Cleveland. In 1952, when I was nine years old, the Cuyahoga River burned furiously. As the river wound its way north from Akron through the industrial Flats section of Cleveland to empty into Lake Erie, it accumulated a toxic miasma of industrial waste. In the century between 1868 and 1969, the Cuyahoga caught fire more than a dozen times. The '69 fire finally triggered the legislation behind the Clean Water Act and the EPA, but the '52 burn was the really big one!

In 1954, my heroes were the Cleveland Indians: Bob Feller, Al Rosen, and Larry Doby. The Indians played before a major league record (since broken) crowd of 86,563 at Cleveland Stadium on Sunday, September 12, 1954, sweeping a doubleheader against the Yankees, 4–1 and 3–2. My dad took me to that September Sunday doubleheader. The wins made the Indians 104–40 on their way to

25 The Robert Hobbs who had introduced me to Scios.

a then American League record of 111–43 and all but ended the chances of the Yankees, who had won the World Series each of the previous five years. The Indians were then swept in the World Series by the New York Giants (Pettica 2013).

The appeal of the clinic was not the weather (particularly compared to Florida), the natural environment, or the faded baseball glories; it was the colleagues, particularly in cardiology. Eric Topol had just become the head of the cardiology group, and Steve Nissen was his second in command. Both had strong ties to the University of Michigan, and both were rising stars in academic cardiology. The cardiac imaging and electrophysiology teams were outstanding, and the interventional cardiology group at the clinic was world-class. In addition, the top-ranked cardiac surgical program drew in a steady stream of complex, challenging patients who required cardiology evaluation and management. It was an exciting, challenging environment.

In those days, the clinic had an Omni hotel on campus, where wealthy patients from Middle Eastern and Arabian Gulf states arrived with a retinue and were known to rent a whole floor for their visits to the doctor. The restaurant at the Omni, called Classics, was a highly rated Cleveland venue. A lunch there was not to be missed, particularly when someone else would pick up the check. A week after the call from California, Robert and I sat down to talk with Dick Brewer. Dick, a tall, athletic-looking executive wearing a well-cut dark suit, looked like a big-time college basketball coach outlining the final two minutes of a critical game.

Leaning forward, he wasted no time getting to the point. “We're in trouble,” he said. He confirmed the failure of the Bayer deal and the fact that Scios would have to do a new phase 3 trial if Natrecor were to remain viable. “I want to hear from people who have actually used Natrecor. Is it a good drug? Is it worth doing another trial?”

Our lunch went on for an hour and a half. It was a typically intense Dick Brewer performance. Almost without a breath, he continued, “I'm visiting as many investigators as I can. I want to know what you think." He followed up with a flood of detailed questions. "What about the dose?"

I answered, "The dosing needs more research. It could be improved—0.015 may be too much."

"What about using a bolus to start?"

Bob answered, "Probably okay, since you have to have an IV, anyway."

"What about the falls in blood pressure? How long did they last? Were they a serious problem?"

We talked this one over. We had both seen reductions in blood pressure, but they weren't a big problem. After all, the half-life of nesiritide was less than twenty minutes.

"Would they pose a problem in patients with coronary disease?"

Again, we talked it over. We had used the drug in patients with heart failure due to coronary disease, and they had not experienced any particular difficulties. And besides, vasodilation, which was the mechanism of action (*MoA*) of the drug, was also MoA of nitroglycerin. Nitro was certainly okay for coronary patients.

Without using or taking notes, Brewer worked his way through every issue that the previous advisory committee had raised and every question that the FDA had about safety. At the end, we had run out of time. Robert and I both had patients scheduled for afternoon office visits, and Dick had a flight to catch. "You'll hear from me soon," he said on his way out the door.

Actually, I don't think either of us ever heard directly from Dick on the question of another trial again. He had the information he wanted; we had enjoyed a good lunch. That was enough for everyone.

We could only infer that the other investigators who Dick visited must have given him similar answers to his list of questions, because the answer to, "Do we call it a day, or raise the money for another trial?" turned out to be "another trial."

What we did hear, quite soon, was that Jim Young, our Cleveland Clinic colleague, would head a new Natrecor trial to be called "Vasodilation in the Management of Acute Congestive heart failure," or VMAC. Young had been recruited to the clinic to head the heart failure program. Among his many other accomplishments, he had played the sousaphone in the University of Kansas marching band. On weekends, he loved riding his Harley. Jim Young was well known in the heart failure community, and he was not shy.

Among the industry giants known as Big Pharma, doing two or more costly phase 3 trials to support approval for a new drug is not uncommon. Sometimes the Big Pharma companies may run multiple trials for more than one indication,[26] and occasionally a sponsor may choose to do two parallel trials for statistical reasons.[27] In contrast, for a struggling biotechnology company like Scios that was basically living hand to mouth, another trial meant another demanding round of fund-raising. Brewer had found the money and the leadership for another trial.

Like the 325-326 trials that Bill Colucci had led before, the VMAC trial had to answer specific regulatory questions on a short timeline and a tight budget. The new study would essentially take all comers. Patients with coronary disease, those with preserved LV function, those with gradual worsening, and those with first-time symptoms were all eligible, as long as they had heart failure with shortness of breath severe enough to require admission to a hospital. This would meet the need for safety data in various different disease states, but with the downside that some might not respond as well as others.

The individual investigator at each site could decide on a case-by-case basis for each patient whether or not to place a Swan-Ganz right heart catheter. The design would allow the investigators to compare the effect of standard care plus one of three possible randomly assigned treatments—nesiritide, nitroglycerin, or placebo—on the absolute measured change in filling pressure (PCWP) between baseline and three hours of treatment in the catheterized patients and

---

[26] As Bayer and J&J did for rivaroxaban, first for prevention of postoperative thrombosis in joint replacement and then for prevention of stroke in atrial fibrillation.

[27] This strategy was used in the EVEREST Trial of tolvaptan in heart failure (Gheorghiade et al. 2005).

on the change in the patients' self-evaluation of dyspnea—shortness of breath—from baseline to three hours in all patients.

To deal with the criticisms of the assessment of dyspnea as done in the 325-326 trials, in VMAC dyspnea and global clinical status were to be "assessed using a nonvalidated symptom scale that is similar to the symptom scale used in a prior nesiritide trial." But this time, the protocol required that "to avoid potential bias, neither the study staff nor the health care team was allowed to discuss or assist the patient in completing the symptom evaluation form" and "in the catheterized stratum, symptom evaluation forms were completed before hemodynamic measurements had been obtained, and hemodynamic results were not discussed within hearing range of the patient" (Publication Committee for VMAC 2002). In addition, multiple secondary measurements and end points would be collected.

VMAC was a very, very complex trial.

Under Jim's leadership, VMAC got under way in October of 1999 and completed enrollment of 498 subjects in August of 2000. Once again, driven by their continued enthusiasm for developing a new and potentially better treatment for acutely ill heart failure patients, the investigators at fifty-five sites in the United States completed the VMAC trial in record time. Like its predecessor 325-326 trials, VMAC had the support of doctors who sincerely hoped that they would help develop something better for their patients.

* * *

The internal timelines for data analysis, preparation, and submission were not as amenable to enthusiasm. Once again, every bit of data and every talking point in the FDA package had to be checked and rechecked. The cardiovascular and renal drugs advisory committee would meet for the second time to review the application for approval of Natrecor on Friday, May 25, 2001.

## Chapter 7

# APPROVAL, NOT UNCONDITIONAL LOVE

The cast of characters for the advisory committee meeting included five committee members who had participated in the 1999 meeting. Milton Packer reprised his role as chairman. Academic cardiologists Tom Graboys, Marv Konstam, Joann Lindenfeld, and Ileana Pina had returned as committee members. Tom Graboys had been a Brigham fellow at the same time I was there and had become a leader in the Boston clinical cardiology community. Marv Konstam, another Bostonian from Tufts who was usually seen with his Red Sox cap folded into a jacket pocket, was the principal outside reviewer. Joann Lindenfeld, from University of Colorado, had voted for Natrecor previously; Ileana Pina, now a professor at Case Western Reserve, had voted against it.

Among the new faces on the committee were Jeff Borer, the chief of cardiology at Cornell in New York City, who was unusually soft-spoken for a Brooklynite, and Steve Nissen, the section head of clinical cardiology from the Cleveland Clinic. Since his undergraduate days, Nissen had cultivated a role as something of a gadfly. He and Eric Topol would subsequently become widely known with the publicity from the Vioxx safety controversy with Merck.

Once again, Darlene led the Scios team, which again included Bill Abraham but also now Jim Young, medical director of the Kaufman Center for Heart Failure at the Cleveland Clinic Foundation and principle investigator for VMAC. Thus, Cleveland Clinic cardiologists were sitting on both sides of the table.

Dr. Ray Lipicky, the director of the FDA cardiorenal division, was also back.

Whether they were representing the sponsor or serving on the advisory committee, all these physicians were at least acquainted, and most of them were friends in the tight leadership circles of academic cardiology.

All of the quoted dialogue in this chapter has been directly reproduced from the transcript of the meeting (US FDA 2001). After the formalities, the meeting began with a distinctly unusual episode of role reversal; Ray Lipicky from the FDA presented the background data on nesiritide on behalf of the sponsor, Scios. His remarks framed the issues clearly and with an unusually long-term perspective. He began, "This … should really start in 1987. That is the last approval date for an intravenous therapy for acute heart failure, and that approval was based entirely on pulmonary capillary wedge pressure change."

He continued, "From 1987 to the present day, there have been enormous changes in what one thinks one ought to know for the approval of something that [will be used] IV [by intravenous administration] in hospital for acute heart failure, and it is during the midst of all of that that the nesiritide program began, and is now culminating."

Then he moved on to summarize the questions that the application raised.

- With a vasodilator drug, was it necessary to decrease blood pressure in order to get a change in pulmonary capillary wedge pressure?
- Was a fixed dose regimen practical?
- The agency needed additional efficacy and safety data, and those data had to include patients with severe illness and with coronary disease; had this requirement been met?
- And, finally, did the data confirm the effect of Natrecor on the symptoms of acute heart failure?

Darlene then took the floor. Her first task hurdle was to explain why the VMAC protocol, in the second phase 3 trial, had specified a dose of Natrecor that had never been previously tested. After the 325-326 studies, Scios had consulted with Dr. Nancy Sambol, a pharmacologist at UCSF (University of California–San Francisco) who had developed computer models to minimize reductions in patients' blood pressure while maintaining the reduction in filling pressures (PCWP); she used the modeling to optimize the dosing. It was a major step forward in technology. As Darlene explained, "We took advantage of the pharmacokinetic and pharmacodynamic profiles of Natrecor that were well characterized in earlier studies at higher doses, and what we did was we evaluated systematically [multiple] potential dosing regimens of Natrecor. This was done by Dr. Nancy Sambol at the University of California–San Francisco, who used a PK/PD model to simulate 24-hour effects of Natrecor on pulmonary capillary wedge pressure and systolic blood pressure, and to do this with candidate regimens of Natrecor."

The modeling approach was a far more sophisticated method than the essentially "back-of-the-envelope" dosing decisions for the Colucci trials, and it had worked.

She then went on to address the trial design. "After agreement with the agency, the primary objectives of the trial were to compare the clinical and hemodynamic effects of Natrecor to placebo when added to standard therapy. I want to emphasize that this comparison to placebo for standard care was primarily to assess efficacy, whereas the comparison to nitroglycerin was primarily built into the trial to study safety."

In response to the questions about severity of illness and coronary disease that were raised during the 1999 advisory committee meeting, and repeated again in Lipicky's summary, she noted that "the study aimed to enroll the sickest patients with decompensated heart failure by limiting enrollment only to those with dyspnea at rest or Class IV symptoms at presentation." In addition, the patients in VMAC were followed for six months after the trial to determine their mortality rates. At the time, these changes from previous studies—a new improved dose, sicker patients, and longer follow-up—all seemed like sound ideas.

In the double-spaced meeting transcript (US FDA 2001), Darlene's full explanation of the VMAC protocol, the document that explains in precise detail how the trial was done, runs to about twenty-four pages. The next forty-plus pages of the transcript cover questions from the committee about the protocol. Looking back on the events in context, from the viewpoint of the Natrecor development team, a single trial designed to address the multiple different questions from the FDA and the advisers made sense. For Scios and for Natrecor, time and money were rapidly running out; there was no other way. On the other hand, a major pharmaceutical company with substantial resources would probably have divided the issues and addressed them in two or three separate studies.

The result of the single-trial approach was that the VMAC data were complicated. Even Darlene and Jim Young would trip over some of the issues.

* * *

Jim Young, who had served as the principal investigator for VMAC, presented the overall trial results. Before he went into the details, he made the same critical point that Darlene had highlighted earlier. He said, "Before specifically reviewing the primary and subsidiary endpoints of VMAC, it is important to remember that this study was designed to demonstrate efficacy when compared to placebo plus standard care." Then he continued, "With safety to be generally assessed by comparing nitroglycerin to Natrecor plus standard care."

Both Jim and Darlene emphasized that the trial design used a placebo comparator for efficacy and an active comparator for safety. It was an unusual approach, but the FDA had approved it. Outside the FDA and the more experienced clinical trialists on the panel, very few individuals would ever grasp this critical issue. The much more common approach is for a single clinical trial to use either a placebo or an active comparator in the control group to determine both safety and efficacy. Placebo controls are often used in trial designs in which all patients receive a standard treatment in addition to either the study drug or placebo. Active comparator designs are used when the study drug being tested is intended to replace standard treatment.[28] Combining both placebo and active comparators in a single trial design, one for efficacy and one for safety, would inevitably cause confusion and misinterpretation about which comparator represented efficacy and which represented safety.

Young then moved on to discuss the results. The protocol that reflects the design of a trial usually declares that one or two *primary efficacy end points* will be the critical measures. These are chosen to answer the scientific questions about whether or not the drug works. They are the most important questions in the trial; they define the benefit against which the safety risks will then be weighed. The biostatisticians who work on the trial design plan their analysis so that it has the most statistical power to analyze the primary efficacy and safety end points.

The VMAC trial had been a success. He began, "The first primary endpoint, mean change in PCWP [the filling pressures measured by the Swan-Ganz catheter] shows significant reduction by Natrecor compared to placebo at all time points. Indeed, the onset of response to Natrecor is rapid, with a significant decrement noted first at fifteen minutes and a peak reduction noted at about one hour, with this change maintained out to the three-hour placebo controlled mark."

Then he went on to the second primary end point, patient-reported dyspnea measured three hours after starting the study drug. "Dyspnea [all patients] improvement with Natrecor was statistically significant compared to placebo."

He then fell into the trap that he had warned the committee about only a few moments earlier. He compared the efficacy of Natrecor to that of nitroglycerin, stating, "In contradistinction to Natrecor, nitroglycerin did not demonstrate statistically significant

---

[28] The recent ROCKET-AF trial, which compared rivaroxaban to warfarin for stroke prevention in patients with atrial fibrillation, is an example of an active comparator trial design.

improvement when compared to placebo." Nitroglycerin had been included as a safety comparator. The 325-326 data had already shown that Natrecor efficacy was comparable to nitroglycerin. Now, the principal investigator had made efficacy comparisons with nitroglycerin, which were not a primary end point in the trial, fair game for the panel and for everyone else.

Early on in the long day, the issues were clear. VMAC efficacy data showed that after three hours of treatment *when used on the background of "standard care" for everyone*, Natrecor as compared to placebo produced a greater reduction in the cardiac filling pressures, and the reduction was statistically significant. In addition, after three hours of treatment *when used on the background of "standard care" for everyone*, Natrecor as compared with placebo was associated with a greater improvement in dyspnea, the primary symptom of acute heart failure, which was also statistically significant.

Most of the remainder of the efficacy discussion would be picking apart the subgroups. At one point, commenting on the lack of statistical power in one of the subgroup analyses, Ray Lipicky said, "So don't blame it on the sponsor. We [the FDA] were part of this, and we did not think that there had to be a plan for how to handle that, and that's our fault."

But soon, some of the discussion became edgier. If the background standard care that all patients received had been state of the art across the board, then overall, the trial had been a success. True to his self-appointed role, Steve Nissen introduced the concern that the quality of the standard care in the trial might actually have been substandard. First, with regard to diuretic dosing before randomization, he said, "But I still think that what's going on here is that investigators are involved in protocols like this. They want the drug to work. They're rooting for you, and I think that may have subtle effects and maybe some not so subtle effects on how they practice medicine." Then, a short time later, discussing the administration of the placebo infusions, Nissen said, "I think that the investigators were really not titrating according to hemodynamics. I think they were just ignoring it because otherwise you would have expected when there was placebo in the bag for those doses to have been escalated. And I must say it tends to confirm my hypothesis here that the investigators were really laying off of standard therapy." This was a harsh and very thinly veiled criticism of the investigators, a first-rate group of academic cardiologists. It could also have been interpreted as a criticism of Nissen's Cleveland Clinic colleague, Jim Young.

No one else commented on Nissen's ideas.

Then the presentation moved on to safety. Just as with efficacy, the primary analysis of safety carries the greatest statistical power. For a new drug like Natrecor, as both Darlene and Jim had emphasized in their presentations early in the day, the primary safety analysis for VMAC was designed to compare Natrecor to nitroglycerin, an active comparator drug with a somewhat similar mechanism of action.

In the 1999 advisory committee discussions, the advisers' concern that Natrecor would cause excessive reduction of blood pressure (*hypotension* in medical terms) had surfaced repeatedly. Darlene could now address that worry with an "oranges-to-oranges" comparison, using the VMAC data that compared the rate of hypotension with Natrecor to the rate of hypotension with nitroglycerin. "During these first twenty four hours, symptomatic hypotension was similarly reported with these two agents, occurring in five percent of nitroglycerin patients and four percent of Natrecor patients." She added, "And, again, there was no adverse event that was reported more commonly with Natrecor." Because reduction in blood pressure can impair kidney (*renal*) function, renal function had been singled out as another important safety measure. There were no differences in renal function between patients randomized to nitroglycerin or Natrecor and no differences in the need for dialysis in the two groups.

She went on to review the safety data for the entire Natrecor development program. The program now included 941 patients who had received the drug. "Overall ... there is no difference in the reporting of any serious adverse event within fourteen days between Natrecor and a variety of control agents."

Finally, she addressed deaths in all the patients who had received Natrecor in the clinical development program to date. "Now, as is usually the case when there are no differences overall in integrated summaries, the point estimates in individual studies will vary. This is exactly what we observed with the four largest trials."

As the number of patients who had gotten the drug grew larger, it became clearer that the overall death rates for patients who got Natrecor and patients who did not were moving toward similar figures. But—and this was a critical qualification—since there were now nine relatively small trials, in some of those trials, more of the Natrecor patients had died, and in other trials, more of the control patients had died.[29] Analyzing the various trials separately would be contrary to everything that the FDA had asked for in the previous advisory committee meeting. In addition, subgroup analyses could bias the data in one direction or another; everyone at the agency and on the committee knew that. Knowing it wouldn't keep it from happening.

As the discussion continued, "big data" lumbered into the room and took a seat. Big data was the eight hundred–pound gorilla at the table. Just a few years ago at the end of phase 2, the goal had been a safety database with five hundred patients who had received treatment with nesiritide. Now, Ray Lipicky spoke about the previous day's meeting in which an FDA advisory committee reviewed a different drug to illustrate the need for consistent safety data from large trials. He said, "Remember yesterday. Okay? We had one randomized trial with 24,000 patients over four years [that settled the safety issue], [after] 381 controlled clinical trials in the drug development database couldn't come anywhere near addressing the problem. *So we're in a different era now.*[30] Okay? We're starting to think differently, and it's important to recognize that, where the safety database that came

[29] "That's how the Law of Large Numbers works: not by balancing out what's already happened, but by diluting it with new data ..." (Ellenberg 2014).

[30] Italics added.

from 380 controlled clinical trials didn't come anywhere close to one randomized, 24,000 patient trial that ran for four years."

He continued, "And I think you have to recognize that we're in this changing paradigm again. I don't see any way out of that, and how soon it's going to be necessary to have those kinds of trials I don't know."

Actually, I think he did know. It was necessary then, in 2001, for Natrecor. At this point, he was either being diplomatic or purposefully vague. At the time, the agency did not have the authority to require post-approval studies; however, here was the warning, loud and clear, from the FDA. He had said, in essence, that the agency might approve Natrecor after the meeting, but the rules of the game were changing. Big data was at the table. The discussion for most of the afternoon had focused on safety, and the way to end that discussion was to do a large randomized safety trial. Period.

* * *

Following the customary agenda, committee voting came late in the afternoon.

Dr. Packer put the question to the committee. "Should nesiritide be approved for the treatment of decompensated heart failure?"

Dr. Pina: "I think with all the caveats that we have stated and with all the statements that we have given, then yes."

Dr. Konstam: "I'm going to vote for approval."

Dr. Artman (committee member): "Yes, it should be approved."

Dr. Graboys: "Approval."

Dr. Borer: "Yes."

Dr. Lindenfeld: "Yes."

Dr. Nissen: "I need to explain my answer here ..." Steve had been a campus activist during his undergraduate days at Michigan. Decades later, I have met older faculty in Ann Arbor who still remember his rhetorical style.

"I think you've shown efficacy unequivocally, and I think it's well done. Obviously it's a very difficult trial, but I think you met all the benchmarks both on wedge pressure and on symptoms, and I think that that was clearly well done.

Therefore, I am prepared to vote for approval, but only if there is a black box warning that says something like this, and I'm not necessarily very good at writing these: 'this agent may produce moderate or severe hypotension at recommended doses, which may be more prolonged, lasting greater than sixty minutes, than typically seen with intravenous nitroglycerin.'" (Nissen now had plausible deniability with regard to his vote, no matter how it was interpreted.)

Dr. D'Agostino (committee member): "Yes."

Dr. Packer: "Let me put my vote in. I would vote for approval ..."

Then Packer continued in response to Steve's request for a warning. Both men were old pros at the FDA. They both knew very well how the FDA worked, and they were almost certainly well aware that advisory committees have no authority to make recommendations about labeling.

Milton said, "So to obviate the need for Steve's warning coming true, a real educational effort needs to be made to make sure that physicians know about this, and maybe that's what we mean about a warning."

Nissen had the last word.

Now, in contrast to his remarks in the morning, he opted to describe the investigators as experts. "Just to comment a little further, I mean I understand, and obviously there is a lot of subtlety in this, but I want to be really clear here and send up this red flag that the investigators in this study were real pros. I mean, I looked at the list of people doing this. They're people who treat congestive heart failure for a living, and they're very good, and they know how to manage, you know, these kinds of patients."

His message about nesiritide, however, was clear. He continued, "And my concern would be it's a lot easier to come back with phase four data that says the drug is safe and take a warning away than it is to wait until there is trouble and add it on later."

The FDA and the committee had sent their messages to the sponsor. The storm-warning flags had been hoisted. The meeting had started at 9:02 a.m., and it ran, with a short break for lunch, until 5:20 p.m. Now the advisory committee meeting was over. The committee had voted overwhelmingly for approval.

Two and a half months later, on August 10, 2001, the approval letter came. Natrecor was a real drug.

## Chapter 8

# DANCING ON THE TABLES

*Scios National Sales Meeting, 2002.*[31]

Kim Hillis, the national sales manager for Scios, was a compact blonde dynamo; she could have been the star shortstop on a championship softball team. Scios had just wrapped up the first national sales meeting after the launch of its new heart failure drug, Natrecor. The closing party for the meeting was not just a business event; it was also a major celebration. The happy scene included a couple of open bars and tables covered with tapas. The popular band Hootie & the Blowfish provided the live music. Carried away by the satisfaction of the moment and the live music, Kim hopped onto one of the outdoor dinner tables and danced. It was the iconic image of a once-in-a-lifetime moment.

People outside the pharmaceutical industry don't understand the magnitude of the launch for a new drug. The launch includes all the work of hiring and training a sales force and then putting together approved promotional materials for the new salespeople and identifying the health professionals and providers they should target with sales visits. A solid launch can set the sales trajectory for years to come. Like a military invasion, the launch process requires enormous efforts in logistics and planning, in recruiting and deploying experienced personnel, and in focusing the maximum

---

[31] I did not attend. I'm only reporting these events from what I was told by those who were on the scene.

effort for the maximum gain. As the national sales manager, Kim was the general who had directed the effort. She had just led the most successful intravenous cardiovascular drug launch ever. She had every reason to celebrate that night.

The sales meeting celebration meant that the drug had succeeded in the marketplace.

At the time, the meeting seemed like the culmination of a long chain of events. It was time for dancing on the tables.

## Chapter 9

# CHANGE, AND MORE CHANGE

The medical community enthusiastically embraced Natrecor as a new approach to heart failure management. On February 13, 2003, Scios announced financial results for the year ending December 31, 2002. The year 2002 represented the first full year on the market for Natrecor (Scios 2003). Total revenues were $111.2 million, with full-year Natrecor net sales of $107.3 million. The press release read, "With nearly 100,000 patients treated with Natrecor in 2002, this makes Natrecor one of the most successful intravenous cardiovascular drug launches ever."

Scios had been a small biotech company for twenty-three years. Without question, its management had dreamed of success, but there was no firm plan in place for the corporate metamorphosis, for the company to become something totally different.

The Natrecor approval meant that Scios had to change. The organization could not just grow into being a larger Bay Area research biotech; it had to become an organization capable of manufacturing and selling its product on a large scale, a pharmaceutical company.

Dick Brewer and his team had done an incredible job of getting Natrecor approved and then launching it. Now, as Brewer knew from his Genentech experience, the company needed a marketing partner. In its early days, Genentech had partnered with Lilly to produce insulin, using the partnership to bridge the transition from a venture-capital-funded, strictly research organization into a pharmaceutical company. Partnering was the fourth, final step in the process to success. The ideal partner for Scios would be a large, established pharmaceutical company with a successful commercial infrastructure. The partner had to be able to support marketing Natrecor to hundreds of hospitals and thousands of physicians across the country. The first partnering attempt with Bayer had failed; this time had to work to build on the success of 2002.

Brewer had engaged several companies in discussions. Then, unexpectedly, a buyout offer appeared. The search for a marketing partnership turned into a bidding contest for Scios itself.

Very few people know the details of what came next, and I am not one of them. There is no public record. The facts are that Dick Brewer had years of high-level experience in biotechnology with Genentech. He must have known that two decades earlier, Johnson & Johnson had declined the opportunity to acquire the fledgling Genentech for $80 million (Hughes 2011). Brewer's counterpart in the negotiations was Joe Scodari, who had started from Youngstown State and had risen to become worldwide chairman of the Johnson & Johnson pharmaceuticals group. Listening to a presentation, Joe was so quick that he would ask the speaker questions about his next, as yet unseen, slide. Scodari must have known the details of J&J's troubled relationship with Amgen, when the partners feuded over the details of marketing EPOGEN and PROCRIT (Binder and Bashe 2008).

Brewer and Scodari were tough, smart pharmaceutical industry veterans. Scios had an approved novel heart failure drug and a very interesting pipeline, with a potential compound for rheumatoid arthritis, a p-38 kinase inhibitor, in development and another compound, a TGF-β inhibitor, in the lab. J&J had missed one major opportunity in biotech when it passed on buying Genentech and missed another when it soured its partnership with Amgen.

Cardiovascular disease, particularly heart failure, was not an area of strength or research interest at J&J. On the other hand, the promising compounds in the Scios pipeline aligned nicely with J&J's strength in arthritis and other inflammatory diseases. The facts suggest that the primary motivation for J&J's interest was the potential of the Scios pipeline, not the commercial success of Natrecor.

Brewer and Scodari made a deal. On February 11, 2003, their deal made the *New York Times*. "Johnson & Johnson said yesterday it had agreed to buy Scios Inc. for about $2.4 billion to acquire its congestive heart failure drug and rights to its experimental arthritis treatment" (*New York Times* 2003).

Under the terms of the J&J acquisition, Scios would become a Johnson & Johnson operating company and would remain on the West Coast. Instead of a partnership focused on developing Natrecor, Scios had a new distant relative in New Jersey focused on pursuing the development of its pipeline.

Over the next year, change erupted at Scios.

A move from Sunnyvale to a new headquarters campus in Fremont was already under way. Then Dick Brewer developed multiple myeloma, a malignant blood disorder (Leuty 2012). Typically, Dick began an aggressive treatment program, but the entire company was stunned.

J&J brought in Jim Mitchell, a Canadian, a ruggedly handsome former hockey player and an experienced career J&J commercial executive, as the new Scios president. Mitchell's strengths lay in sales and marketing, and initially the commercial side of the business flourished under his leadership.

The good news was that, in the status of an independent J&J operating company, Scios retained its small California biotech feel. The offices were still dog-friendly, and the volleyball pit saw daily use.

The bad news was that roles and responsibilities that had been tacitly understood and accepted in the biotech days became thoroughly scrambled in the course of all the changes. With the commercial success of Natrecor, new clinical studies to expand the indication were on the drawing boards, but the research effort lacked a clear scientific focus. The excitement of building a multimillion-dollar market for the new product dominated leadership thinking, but the market research that would have focused the marketing effort toward emergency medicine, emphasizing simplicity and safety, was not there. Most important, the prospect of going back to do a large and costly phase 4 safety study in heart failure simply to reassure a few skeptics had no appeal; it seemed like a dusty relic from the distant past. Everyone thought that the new studies in planning would be adequate to expand the safety data.

## Chapter 10

# THE BIGGER THEY ARE

Under the acquisition terms, Scios had become a Johnson & Johnson (J&J) operating company, which substantially changed the scale of its financial reporting. The J&J annual report for 2004 listed worldwide total sales of $47.3 *billion* with a worldwide operating profit of $12.8 *billion*. As a matter of corporate policy, J&J does not report financial results from its individual operating companies. Nonetheless, on the scale of J&J's corporate finances, Natrecor sales now amounted to little more than a rounding error.

After almost a decade of cautious enthusiasm about nesiritide, 2004 was a year of simmering controversy in the heart failure community. The mood had shifted from early excitement about the new drug that had succeeded in clinical trials to concern about how the drug was being used. Both the cost of Natrecor, at about $500 per vial (enough drug to treat an average-size adult for a day) and the terms of its commercial success seemed to stir much of the discontent.

On the development front, Abbott brought a new European drug, levosimendan, into US development in phase 3 clinical trials. Levosimendan was a potent contraction enhancer (*positive inotrope*) like the generics, dobutamine and milrinone. Abbott presented it to the cardiology world with some fanfare as a "calcium sensitizer."[32] Levosimendan was also an intravenous compound and thus was a potential head-to-head competitor to Natrecor in the acute heart failure space. Abbott had named a familiar figure, Milton Packer, as the academic leader for the clinical development program. After chairing both of the nesiritide advisory committees, Milton knew the scientific and regulatory issues surrounding new heart failure drug development better than anyone else outside the FDA. Would he be able to help Abbott develop an approvable package? If anyone could knock Scios and Natrecor off their perch, Milton could.

Meanwhile, in the practical world of treating patients, cardiologists and primary care doctors who cared for patients with heart failure felt increasingly pressured by finances. The business side of medicine had reared its ugly head. Medicare, the insurer for most heart failure patients, was spending almost one out of every five dollars on heart failure hospitalizations. The billing rules, and the scrutiny of practices, were becoming more restrictive. Hospitals and payers steadily increased administrative and financial pressures on physicians to shorten length of stay and reduce costs for heart failure admissions, while offering little support for the provision of routine outpatient care.

Physicians and nurses who focused their practices on heart failure identified a substantial number of patients who required repeated hospital admissions, often via the emergency department. They dubbed them "*frequent flyers.*" The frequent flyers, they believed, were the cost drivers for the health care system.[33] The problem seemed straightforward: find a way to manage them without repeated hospitalizations.

---

[32] With apologies to Miss Piggy and the Muppets, the "calcium sensitizer" label turned out to be a good example of "lipstick on a pig." Levosimendan had the same profile of adverse effects as other positive inotropes. Milton and Abbott delayed peer-reviewed publication of the trial results for almost a decade (Packer et al. 2013). Abbott never presented the drug for FDA review.

[33] Actually, few data supported this concept. Although I initially agreed with the clinicians, I later argued that the idea was a product of survivorship bias (Mills 2009).

At the time, many practicing physicians were still familiar with an older—albeit totally unproven—idea for managing severe heart failure called the "*dobutamine holiday.*" The concept was simple and, on the surface, rather appealing. The holiday would give the failing heart some much-needed relief by supporting it with dobutamine, a potent intravenous positive inotrope, for a few days. The patient would feel better, clear his or her accumulating fluid overload, and "re-compensate" for a few weeks. At least, that was how the theory went.

There were, however, a number of problems with dobutamine holidays. Dobutamine caused serious heart rhythm disturbances in some patients, and the patients most likely to experience problems were those with heart failure. Accordingly, most hospitals restricted its use to special nursing units equipped with continuous electrocardiographic monitoring. Second, some small observational studies had suggested that positive inotropes like dobutamine actually increased the long-term mortality (death) rate in heart failure patients. Third, for Medicare, even a short admission counted as an admission; an inpatient stay for a dobutamine holiday did not reduce the financial pressures.

Under the circumstances, practitioners desperately needed a new approach. They wanted a drug that helped heart failure patients feel better and that didn't cause rhythm problems. Working with Scios, Andy Burger and his colleagues had already published data showing that, in sharp contrast to dobutamine, Natrecor did not provoke arrhythmias (Burger et al. 2001). If practicing cardiologists could convince their hospitals to dedicate clinic space and nursing staff in order to develop and run an outpatient Natrecor infusion program, then they could treat the frequent flyers as outpatients with Natrecor and hope to prevent their readmissions. It would be like an outpatient dobutamine holiday program, but with a safer drug.

Thanks to the development and widespread acceptance of outpatient cancer chemotherapy, insurance companies had put in place favorable billing codes for intravenous (IV) drugs given as outpatient treatment. Reimbursement was much better, in fact, for an IV treatment in the clinic than for a routine outpatient visit in the office. The outpatient heart failure infusion clinic looked like a win-win proposition. The patients would get regular IV Natrecor and stay out of the hospital, and the clinic would be financially viable. What could go wrong?

The answer to that question turned out to be long, complicated, and painful.

First, as heart transplantation evolved into a widely accepted procedure, academic heart failure and transplant programs had developed in about one hundred university medical centers around the United States. The physicians heading these programs were the regional and national key opinion leaders of the heart failure community. The academic programs they led depended on referrals of challenging patients from a wide base of community cardiologists for their very existence.

If outpatient clinics at community hospitals could succeed in keeping the difficult heart failure patients at home and feeling reasonably well, these specialized programs would lose referrals. If they had no referrals, they would dry up and blow away.

Some of the academic specialists had worked hard to develop their relationships with the community physicians. As consultants, they knew, respected, and supported their referral base and did not feel threatened by the infusion clinics. On the other hand, some of the less politically adept "key opinion leaders" who were important figures in the academic community began to see Natrecor as a threat to their programs.[34]

Second, and more importantly, the payers caught on quickly. (They usually do, or they don't stay in business!) As charges for outpatient infusions of Natrecor grew, insurance carriers began to seriously question the billings.[35]

---

[34] Putting the onus on Natrecor for this problem seems to me like blaming the baseball for breaking the garage window.

[35] Although there are few data in the public domain and there are widely differing opinions, everything I learned later as an "insider" supports the position that the vast majority of the Natrecor sales in 2004 went to hospital pharmacies. Whatever fraction of Natrecor sales eventually went to outpatient administration (probably about 15 percent), it was almost certainly given in hospital-sponsored clinics.

Finally, and most critically, a vocal segment of the medical community realized that there were no data to support what was, in effect, off-label treatment: the use of an approved acute care (short-term) drug for the regularly scheduled treatment of a chronic (long-term) problem.

Much to her credit, Darlene had seen this problem coming almost three years earlier. In late 2001, she had put together a pilot study, a modest clinical trial called FUSION I. As stated in the description of the trial posted on the research website clinicaltrials.gov,[36] the purpose of the trial was "to assess the safety and tolerability of two doses of nesiritide ... when administered serially as a treatment to outpatients with worsening congestive heart failure ... who are receiving their usual cardiac medications and are at high risk for hospitalization" (Scios 2005a).

For academic leadership, she had turned to Clyde Yancy. Clyde's star was rapidly rising. Gracious, brilliant, a polished speaker, and deeply interested in clinical research, he brought scientific credibility to the question. In the three-month pilot trial, 210 patients were randomly assigned to usual care only or to usual care and a weekly Natrecor infusion. The results were published in September 2004 in the *American Journal of Cardiology*. Overall, the weekly Natrecor infusions made no difference in outcomes (Yancy 2004). The authors carefully worded their conclusions: "Although there were no statistically significant differences among groups by outcome, prospectively defined higher risk subgroups demonstrated significant decreases in cardiovascular events. These results demonstrate the safety and feasibility of administering nesiritide in an outpatient setting. Additional studies are needed to determine the effect of outpatient serial infusions of nesiritide on rates of morbidity and mortality in advanced heart failure." Because there were some hints of benefit in "high-risk" patients, a second trial called FUSION II was planned to follow on. It would turn out to be far too little and far too late.

With the publication of the FUSION I results, both the economic and medical critics of regularly scheduled outpatient infusions had all

[36] https://clinicaltrials.gov/ct2/manage-recs/fdaaa.

the ammunition they needed to push for closing the programs: high cost, no FDA-approved indication,[37] and no evidence of benefits in a randomized controlled trial.

As 2004 ended, the world's attention focused on a humanitarian disaster caused by an enormous tsunami in the Indian Ocean. Without warning, a shift of tectonic plates in response to forces far undersea created widespread effects that impacted everyday life for hundreds of thousands of people. At Scios, something similar happened in microcosm. The transition from the vibrant lifestyle of a small entrepreneurial biotech to the new and far more complex corporate culture as a J&J operating company seemed to have stabilized the leadership and finances. Under the surface, however, widespread resentment was building about the company's failure to launch a large safety study, and in addition, serious opposition to the company's tacit tolerance of infusion clinics was growing. Under the calm surface, the stresses were accumulating.

In the business world, managers like to use what the books call a SWOT analysis. This process is designed to produce a list of *strengths*, *weaknesses*, *opportunities*, and *threats*. SWOT analyses play an important role in the process of annual planning for budgets and personnel. As 2005 began, one of the glaring weaknesses in the SWOT analysis of the Scios organization was the absence of a cardiologist who had practical experience with Natrecor.

The year 2005 was something of a milestone for me. I had completed my formal cardiology training in 1975. By 2005, I had practiced clinical medicine for thirty years: three decades of night and weekend call, of weekends catching up on dictation, and of rounding in the ICU. I had often wondered what working in industry would be like, and I was ready for a change.

The hiring process with Scios and J&J went smoothly. After all, I had been involved with Natrecor since the phase 2 dosing study

---

[37] Physicians have the option to, and often do, use drugs for conditions other than the FDA-approved indication. This factor, taken alone, would not have been a major issue were it not for the high cost and lack of benefit.

and had maintained good relationships with Scios management. On March 25, 2005, I accepted a position at Scios as senior director of medical affairs, reporting to Randall Kaye, MD, the VP of medical affairs. Relocation to California would take some time; I agreed to start on June 20.

* * *

The first wave of criticism hit Scios on March 29, 2005, when *Circulation*, the prestigious journal of the American Heart Association, published an article titled "Risk of worsening renal function with nesiritide in patients with acutely decompensated heart failure." The authors were Jonathan Sackner-Bernstein, Hal Skopicki, and Keith Aaronson (Sackner-Bernstein, Skopicki, and Aaronson 2005). They had selected just five of the nine available nesiritide studies for their analysis.

On April 20, the even more widely circulated and highly regarded *Journal of the American Medical Association* (*JAMA*) published "Short-term risk of death after treatment with nesiritide for decompensated heart failure: a pooled analysis of randomized controlled trials." For this paper, the authors were Sackner-Bernstein, Marcin Kowalski, Marshal Fox, and Keith Aaronson (Sackner-Bernstein et al. 2005). Only three of the available studies were included in the "pooled analysis," which limited the numbers to 485 nesiritide-treated and 377 control patients.

In both papers, the authors had selected data from studies that had already been reviewed by the FDA. They based their conclusions on fewer patients than the FDA had required in the safety data set six years earlier, and they egregiously violated the concept of using all the available safety data for analysis.

The tidal wave of publicity began in earnest in the lay press. The *Circulation* paper was available online on March 21, and on March 22, the *New York Times* carried an article headlined "Popular Heart Drug Impairs Kidney Action, Study Says." Stephanie Saul, the reporter, introduced Steve Nissen by stating that "he cast the only vote against it [Natrecor] while serving on an FDA advisory panel." She went on to say, "his concerns were based on data showing excessive renal problems" (Saul 2005d).

The *Times* responded quickly to the second paper as well. On April 26, another article by Stephanie Saul ran under the title "Johnson & Johnson Adds Data on Deaths to Label on Heart Treatment." In the second paragraph, she covered the facts. The FDA-approved label change was not a warning. It included the data that in clinical trials, 5.3 percent of the patients treated with Natrecor died, compared with 4.3 percent in control groups, and included, "But the new label notes that the data might not be statistically significant because of the small numbers of patients involved."

Then she continued, "Part of Natrecor's sales volume is related to its increased use in outpatient clinics …" and added, "A well-known Cleveland Clinic cardiologist, Dr. Steven E. Nissen, said yesterday that treating patients with the drug in ambulatory settings was 'inappropriate and cannot be recommended.' Dr. Nissen cast the only vote against the product when an FDA advisory panel recommended it in 2001" (Saul 2005b).

Within days, the high-profile media coverage had accomplished three things.

— First, both the lay press and the medical community at large uncritically accepted the unfavorable safety-related conclusions of the Sackner-Bernstein and Aaronson papers in *Circulation* and *JAMA*. The combination of the initial publication in high-profile journals and the *New York Times* coverage had given the allegations instant validity.

— Second, Saul's reporting had firmly and inextricably linked the safety questions to the outpatient infusion controversy.

— Third, in the public's view, her use of the quotations from Steve Nissen had located the center of the opposition to Natrecor at the Cleveland Clinic,

> the *US News and World Report* national number-one-ranked cardiac center (Cleveland Clinic 2015).

The shouts of panic completely drowned out the few quiet voices that dared to wonder aloud, “Is this a false alarm?”

Without question, Ray Lipicky’s warning had come home to roost. A total of 921 actively treated patients at approval constituted a very small safety dataset. Because the patients had been treated in several different trials, the possibility of subdividing the safety data using selective approaches had always existed. In the scientific world, this approach is called a *retrospective ad hoc subgroup analysis*, and the approach is generally suspect because these analyses are statistically shaky at best. The most appropriate, scientifically valid way to answer the safety question, as Ray and many others had said, was to do a large prospective randomized double-blind trial. No matter how urgent, that would now take years. The immediate consequences that the ensuing Natrecor safety panic engendered, like those of a shout of “*Fire!*” in a crowded theater, were disastrous.

## Chapter 11

# THE WISE MEN MEET

I had been on a working visit to Boston as a Scios consultant when the initial Sackner-Bernstein paper was published in *Circulation.* The response of the medical community there was that the publication was a disaster for Scios. Yet when I called Darlene urgently that dreary, wet morning, she was confident that there was no reason for concern. She had been in touch with Jim Young, and they were confident that the medical community would recognize the flawed methodology in the paper. Even after the second salvo, the *JAMA* paper, made it obvious that Natrecor was under a concerted attack, Scios did not mount a public corporate response.

Finally, in April of 2005, Darlene made a public announcement in a press release. She said that Scios would convene a panel of external experts "to review and assess important data associated with its acute heart failure treatment Natrecor." Dr. Eugene Braunwald agreed to chair the panel. Braunwald, the academic leader of the cardiology world, was the distinguished Hersey professor of medicine at Harvard Medical School. The panel would meet in June. In any event, I had no input into the corporate thinking that led to the "Braunwald panel."

Obviously, the company leadership hoped that the panel would look at the data from *Circulation* and *JAMA* papers and also at the safety data presented at the 2001 advisory committee and in the hands of the FDA. Management hoped that the panel would conclude, reasonably, that the authors had used flawed methods, retrospective subgroup analyses, in the publications.

Even though the company leadership had chosen Gene Braunwald, the most widely respected cardiologist of his generation, they did not know him well. He would do what he had to do, review the data, which had already been subject to detailed examination at the FDA, and then tell the company how to end the controversy.

Braunwald arrived at the Peter Bent Brigham Hospital just as I was completing the first year of my two-year cardiology fellowship. Four decades later, I still clearly recall a noontime "cath conference" that Lew Dexter's cardiology section had scheduled to review some unusual hemodynamic findings in a patient with hypertrophic cardiomyopathy. Dexter had invited Braunwald to attend and discuss the case, since Braunwald was a widely acknowledged expert on the condition. Dexter was low-key and gentlemanly; his conferences never started on time. One of Dexter's fellows had put all the patient's hemodynamic data, including the "pulmonary capillary wedge" (see footnote 19) on the conference room chalkboard. Braunwald arrived at 11:59 to find only two or three people, myself included, in the conference room. He muttered a brief greeting and then sat down and intently studied the numbers. When Dexter and the rest of the attendees ambled in at about 12:05, Braunwald stood up, said, "You need to do a trans-septal!" and left.

That was Gene Braunwald. On time, on point, and devastatingly accurate.[38] (Actually, for the cardiologists in training, figuring out *why* Braunwald had said what he did was far better than being told!)

On June 8, 2005, the Natrecor Advisory Panel met in Boston. Dr. Braunwald had handpicked the members:

— John Burnett from the Mayo, a world expert on natriuretic peptides
— Bill Colucci, who led the Scios 325-326 studies
— Barry Massie, a heart failure specialist from UCSF
— John McMurray, a cardiologist from the University of Glasgow

---

[38] The issue in question was whether hypertrophic cardiomyopathy could obstruct the flow of blood from the left atrium to the left ventricle. A trans-septal catheterization would permit direct measurement of the pressures in both chambers, thus eliminating the assumptions involved in Dexter's "pulmonary capillary wedge."

— Chris O'Connor, a leading clinical researcher from Duke
— Milton Packer, the chair of both Natrecor FDA advisory committees
— Ileana Pina, a member of both Natrecor FDA advisory committees
— Bert Pitt, a highly respected senior cardiologist and clinical researcher from the University of Michigan
— Jim Young, the academic leader for the VMAC study.

The panel focused on three major points: kidney function, deaths (*mortality*), and their recommendations.

First, they noted that in the VMAC trial, serum creatinine (a measure of kidney function) increased more often in nesiritide-treated patients than in controls; 8 percent versus 7 percent at five days and 28 percent versus 21 percent at thirty days. However, they commented, "Most of these increases occurred days after discontinuation of the drug."

Second, they said, "Use of nesiritide was associated with a trend toward an increase in mortality at 30 days." They followed this with a detailed disclaimer: "However, the confidence intervals[39] around this ratio are wide and the number of deaths in a pooled analysis of all six of the controlled clinical trials is insufficient to identify or exclude, with confidence, a moderate excess of risk to survival."

Third, the panelists wanted more data. They recommended a trial with several thousand patients "to assess further the benefits and risks of nesiritide compared to standard therapy." And they wanted it to "be initiated without delay" (Braunwald 2005). The Braunwald panel gave the same guidance to the new Johnson & Johnson management that the original Scios team had heard from Ray Lipicky: do a safety study, a big one, and do it now.

To the small number of readers in polite academic circles who could read between the lines, it was clear that the wise men of the panel did not accept either the methods or the conclusions of the Sackner-Bernstein papers. However, they would neither refer to

---

[39] Confidence intervals are a measure of the statistical certainty regarding a measurement. The wider the confidence intervals, the more uncertain the exact measurement is, and vice versa.

the publications in their press release nor publicly second-guess their friends and associates, the respected editors and reviewers of *Circulation* and *JAMA*. They were also not going to exonerate Natrecor after the original call for a safety study had fallen on deaf ears. Someone had shouted "*Fire!*" but the Braunwald panel was not the fire department. In both the lay press and the medical world, the safety worries about Natrecor were still unresolved.

Addressing the public controversy, the panel's three final recommendations included limiting the use of nesiritide to patients hospitalized with acutely decompensated heart failure. This was the indication based on the VMAC study population and the indication the 2001 advisory committee had discussed. Second, they recommended that nesiritide should not be used for intermittent outpatient infusion or scheduled repetitive use. Third, they asked that Scios should proactively educate physicians about when nesiritide should and should not be used (Braunwald 2005).

For Scios, these recommendations meant that the passive approach of saying nothing about off-label use would no longer work. Henceforth, the company would actively discourage intermittent scheduled outpatient infusions of Natrecor.

Following the advisory panel meeting, in July 2005, both the weather and the language grew more heated. Eric Topol, head of the Cleveland Clinic's department of cardiovascular medicine, added his voice to the public outcry with a "perspective" opinion piece in yet another high-profile medical journal, the prestigious *New England Journal of Medicine.* Topol's paper was titled "Nesiritide—not verified" (Topol 2005). Topol had achieved academic rock star status for understanding the importance of the big data that large-scale randomized trials in interventional cardiology delivered and for conducting very large trials successfully. He wrote prolifically, spoke at session after session at the national meetings, and had already published a second edition of his own *Textbook of Cardiovascular Medicine* (Topol 2002), a direct competitor to *Braunwald's Heart Disease: A Textbook of Cardiovascular Medicine* (Braunwald 2004). He rapidly rose to tenured professor status at the University of Michigan, and at age thirty-six, he assumed the chairmanship of cardiovascular medicine at the Cleveland Clinic.

Topol had already played a prominent role in the public controversy over the safety issues surrounding Vioxx, Merck's oral anti-inflammatory and analgesic drug (Topol 2004). Now he spoke out about Natrecor as well. He wrote, "How can a drug that is associated with higher rates of both renal dysfunction and death than placebo—and that costs 50 times as much as standard therapies and for which there are no meaningful data on relevant clinical end points—be given to more than 600,000 patients and be promoted throughout the United States for serial outpatient use, an indication not listed on the label?" In a single carefully crafted rhetorical question, he had given the Sackner-Bernstein and Aaronson conclusions status as fact, cast doubt on the VMAC trial, and raised the question of improper marketing practices (Topol 2005).

On July 20, less than a week later, J&J announced in a press release that Scios had "received a subpoena from the US Attorney's office in Boston, requesting documents related to the sales and marketing of the company's heart failure drug Natrecor." The following day, J&J announced that it would buy back the European marketing rights to Natrecor from GlaxoSmithKline (GSK), thus terminating a potential partnership in the EU that had antedated the J&J acquisition of Scios. Meanwhile, the media coverage related to Scios and Natrecor continued to intensify. Stephanie Saul's July 21 article in the *New York Times,* "U.S. Looking at Marketing by Johnson & Johnson," reported on the subpoena and the GSK buyback and then devoted six paragraphs to Topol's *New England Journal* piece and the Sackner-Bernstein and Aaronson papers (Saul 2005e). Both federal legal questions and business marketing practices had been tossed into the stew of controversy surrounding the Scios and Natrecor.

* * *

Very quietly, over the next few months, J&J conducted its own internal review of the allegations of Natrecor-related renal dysfunction and increased mortality. As a Scios VP, I participated firsthand in the process. Because the internal review committee members were senior J&J executives, many of whom were physicians with long careers that included both regulatory and industry experience, the data review and preparation for the J&J meeting were more intense than for any FDA presentation. This review was not just about Natrecor; our careers and Scios itself were at stake. In the actual review committee meeting, the questioning from the committee was harsh, detailed, and unrelenting. The generally polite decorum that was de rigueur at the FDA advisory committee on the public record did not extend to the internal review.

Once again, the internal committee reviewed all the safety and efficacy data that the FDA and the Braunwald committee had seen, as well as the Sackner-Bernstein publications. The questioning from the committee was not the polite peer-to-peer dialogue of a public hearing; it was the bosses calling their subordinates on the carpet. Although it was never made public, the internal review was the most intense scrutiny of the Natrecor development program data ever conducted. Much of the intensity was driven by the financial magnitude of the decisions hinged on the findings. Whatever the reviewers decided, millions of dollars, in either income or expenditures or both, were at stake. At the end of the process, the internal reviewers agreed that the published analyses from Sackner-Bernstein and his colleagues were fundamentally flawed and that their conclusions were not justified.

Under the circumstances, the only way to put an end to the safety controversy would be to conduct the large-scale randomized trial that the Braunwald committee had recommended. The trial would cost millions of dollars, and it came at a time when Natrecor sales were plummeting. On the other hand, Johnson & Johnson's reputation for corporate integrity was at stake.

* * *

As the California summer gently turned into autumn of 2005, the sales and marketing team at Scios turned their considerable skills to the quixotic task of developing advertising and education programs that might have some chance of countering the impact of the repeated blows that Natrecor had experienced. They developed "the recommended use campaign." The medical affairs group supported educational programs that were strictly monitored for appropriate content. But without new data, commercially, Natrecor really did not have a chance.

In my area, medical affairs, the Acute Decompensated Heart Failure National Registry (ADHERE)[40] was the first to go. Over the years, Scios had invested roughly $40 million in the registry. The decision was made late in 2005; the public announcement came on March 9, 2006. Many prominent academic physicians, both cardiologists and emergency medicine specialists, had participated in the growth and development of the registry: Tom Heywood from Loma Linda, Greg Fonarow from UCLA, Clyde Yancy from Baylor, Adrian Hernandez from Duke, and Frank Peacock and Charles Emerson from the emergency medicine group at Cleveland Clinic, among many others. Largely based on their publications reporting ADHERE registry epidemiologic data on over a hundred thousand hospital admissions for heart failure, the concept of heart failure had expanded to include heart failure with preserved systolic function, and the importance of kidney function in determining heart failure outcomes had been documented. ADHERE had made a huge scientific contribution.

ADHERE, however, had important design flaws. Because the data entry was based on unique hospital admissions rather than individual patients, researchers could not follow an individual patient's outcomes.[41] In addition, the vast number of hospital admissions included in the database had grown so large and unwieldy that adding more information would no longer impact the conclusions. ADHERE had morphed from a PT boat into an aircraft carrier. Even though the registry was in my area of responsibility, I fully supported shutting down the data-acquisition process.

When the discussions turned to cutting actual clinical research and development trials, they grew more heated as the limited options

40 The ADHERE registry collected information on unique hospital admissions for heart failure.

41 If we had been able to track individuals over time, ADHERE would have provided the safety data we so desperately needed.

became clearer. For the clinical research group at Scios, the prospect of planning a new large, randomized trial raised multiple internal conflicts. Where would they find the money and the human resources for an additional major US safety trial? Discussions were already under way with the European regulatory agency about the design of the ETNA trial, a study that management hoped would provide the data required to get approval for the marketing of Natrecor in Europe. As they looked at options, FUSION II, the follow-on outpatient infusion study with some nine hundred subjects, was already under way. It had a projected completion date in the second half of 2006. There was nothing to be gained by stopping the trial early. Most of the costs were already sunk, as the finance team pointed out.

A previous three hundred–patient randomized pilot trial called NAPA had studied nesiritide infusions during coronary bypass surgery to protect kidney function. The results looked favorable, and the commercial team wanted to do a larger label-enabling trial as soon as possible. In addition, a small phase 2 study looking at Natrecor in patients awaiting heart transplant about to start. Natrecor sales were going steadily down. Some programs were going to be cut; who would J&J have to throw out of the sinking lifeboat?

The options rapidly came down to a choice between trying to dodge the safety trial bullet by combining safety data from the multiple smaller studies the company favored or cutting everything else to do the recommended large safety trial. Jim Mitchell, Randy St. Laurent—who was Jim's aide-de-camp and overall strategist for relations with the medical community—and I were tasked with discussing the options with Milton Packer.

If Packer, with his expertise in trial design, could be persuaded to support building a safety database from the multiple smaller trials approach, the decision would be much easier and far less painful. By this time, Milton had moved from Columbia to the University of Texas Southwestern Medical Center. We met there and commandeered a vacant classroom so we could outline options on the blackboard. It was hot (no surprise in Texas), and as the discussion went on, the climate in the room grew perceptibly warmer. Finally, Milton said, "Look, you guys [J&J] promised us [the academic community] a beautiful, new fifty-four-inch color TV, and we still haven't gotten it. Now you're asking us to accept nine eighteen-inch black-and-whites and call it okay because the screens will add up to the same area. No, it's *not* okay!"

Milton's heated outburst answered the lifeboat question. Everything, including the ETNA study in Europe, the NAPA study in coronary surgery, and the transplant study, would go overboard. There was only going to be one study.

* * *

J&J is a very large company, over one hundred thousand employees, with enormous clinical research talent and depth. It is not, however, nimble. Getting consensus to approve and fund a multimillion-dollar study takes some time. In addition, upper management vastly complicated the study planning by insisting on trying to salvage potential approval of Natrecor in the EU by making the study an international project with efficacy as well as safety end points. It was not a good decision.

On June 1, 2006, just a few days short of a year after Braunwald and the panel met in Boston, Scios put out a press release that announced its intention to fund an international outcomes study of some seven thousand patients "to further assess the benefits and safety profile of Natrecor in patients with acutely decompensated heart failure." The outcomes study, the release continued, "will be the largest and most comprehensive study of its kind."

The announcement put the critical decision in the public domain. J&J would make the investment to defend its acquisition.

* * *

Immediately, inside Scios, the critical study-related issues requiring decisions started to pile one on top of another, and the pressure to make those decisions mounted. In the press release, the company had committed to actually starting patient recruitment for the new trial in the first quarter of 2007.

Two strong arguments quickly excluded the Scios research group from running the trial from within the company. First, Scios did not have an internal research and development organization with the size, scale, and experience required to conduct a seven thousand–patient international trial. Second, unless Scios appointed an independent academic research organization (ARO) to oversee the entire process, the skeptics would never accept the data. The first step toward actually doing the trial had to be finding an ARO partner.

The list of North American[42] AROs with the capability of handling a seven thousand–patient trial was very short. The TIMI Group, based at Harvard, had done many large trials in acute myocardial infarction and could handle the logistics; however, TIMI was not in the running. Gene Braunwald had played a critical role in the development and leadership of the TIMI Group. Even if he somehow recused himself from the operation of the trial, the public perception of "his ARO" running the trial would not be favorable. The other two large organizations that remained under consideration were the Population Health Research Institute (PHRI) at McMaster University in Hamilton, Ontario, Canada, headed by Salim Yusuf, and the Duke Clinical Research Institute (DCRI), at Duke University in Durham, North Carolina, led by Rob Califf.

Both AROs had strong reputations, and both had articulate and enthusiastic supporters on the decision-making team. Without a doubt, the choice of the ARO would have a major impact on the design and execution of the trial. Yusuf, who had been a Rhodes Scholar at Oxford and had also spent time at the US National Institutes of Health, had a well-deserved reputation for innovative trial designs. In contrast, with the exception of his medical residency at UCSF, Califf had been at Duke for his undergraduate, medical school, and cardiology training and for his entire subsequent medical career. He had been a lead investigator for numerous trials; most importantly, under his leadership, DCRI had set the standard for ARO independence from industry sponsors.

---

42 Natrecor was essentially a North American product, with very limited regulatory approvals outside the United States.

At a level far higher than mine, the nod finally went to DCRI.

* * *

In 2006, as the organizational consequences of the Braunwald panel recommendations played out, Darlene, the last of the original Scios cast, left the stage. She had been a constant throughout the development of Natrecor. The R&D segment of the company had no plan for succession at her level.

Meanwhile, the DCRI leadership needed answers to critical questions. How many patients? What end points? What entry criteria? What comparators? What statistical methods? Within the existing J&J cardiovascular research and development organization, corporate expertise focused on antithrombotic agents, not heart failure. The combination of DCRI's policy of operational independence from corporate sponsors and the absence of an empowered internal R&D leader with years of Natrecor experience at Janssen, the J&J pharmaceutical division, meant that no strong voice spoke up to represent the strategic interests of the company in the debates about those critical operational questions.

Adding to the internal R&D problems, Darlene's strategy of involving one expert after another—including, over the years, Bob Hobbs, Bill Colucci, Bill Abraham, Jim Young, and Frank Peacock—had engaged a number of key individuals in academic heart failure over the years; however, the downside to her approach was that no single person in the academic medical community had acquired long-term experience with nesiritide that included deep familiarity with the regulatory issues and with the safety data. The consequence was that none of the former trial leaders could step into the DCRI discussions to represent the company.

In my view, the final and critically important planning error was that the concept of reversion to the mean[43] in acute heart failure treatment was simply left out of the discussions about the protocol

[43] See chapter 17 in Kahneman's *Thinking, Fast and Slow* (Kahneman 2011) for a superb discussion of this concept.

for the new trial. Typically, after coming to the hospital, if a heart failure patient is going to improve, he or she will get better with standard treatment over the first day or two. At the end of twenty-four to forty-eight hours, he or she will be breathing with less effort and resting in bed and sitting up in a chair more comfortably. As both the Colucci 325-326 and the VMAC studies showed, *the only real advantages of Natrecor* were that (a) patients receiving it would be more comfortable a few hours earlier while (b) getting a drug that was less dangerous than dobutamine or milrinone, more easily administered than nitroprusside, and better tolerated than nitroglycerin.

If the trial had been designed only for safety end points, this would not have been an issue. On the other hand, with upper management's decision to fight a two-front war and include efficacy end points, designing a trial to highlight early and less complicated treatment was critical. If the new study included both safety and efficacy and then showed safety but failed to reconfirm efficacy, then Natrecor would be lost. (Another alternative, that Salim Yusuf at PHRI had favored, would have been to test efficacy for an entirely different indication while collecting the safety data, thus hedging the bet. It was a good idea.)

What both Scios and DCRI failed to factor into the trial design was that the timeline of an individual patient's symptomatic response to treatment depends on many other factors in addition to the elevated pressures in his or her lungs. Whether or not he or she has concomitant lung disease, how long the pulmonary pressures have been elevated and how successfully the individual patient has adapted to his or her condition influence both the nature and severity of symptoms. In the face of reversion to the mean, convincingly demonstrating rapid early improvement in patient-reported symptoms with Natrecor would require not only a large number of patients, but also evaluation and randomization into a trial within an hour or two after presentation for care.

Scios had conducted discussions with teams of academic cardiologists and emergency medicine specialists about the feasibility of an early enrollment heart failure trial as early as 2005. Both TIMI and PHRI had completed "mega-trials" that had enrolled thousands of patients with acute myocardial infarction. The success of those trials showed that very early enrollment could be done. Yet the final protocol would not require early enrollment. In fact, patients could be enrolled up to twenty-four hours after their first dose of IV diuretic.

As if the protocol-sanctioned delay in enrollment were not enough of a handicap, the reversion to the mean problem was compounded by the decision to change the measure of how well Natrecor worked (the *efficacy end point*) from improvement in dyspnea at three hours to improvement in dyspnea at six hours after the start of the study drug. The combination of delayed enrollment and delayed assessment, at least from my viewpoint, meant that the chances of showing rapid early improvement with Natrecor would be substantially reduced. The patients in both the Natrecor and control groups would have a day of treatment before the end point assessment. Regardless of what treatment they had received, if patients were going to improve, they would do so before the end point assessment.

To be fair, the trial design involved important trade-offs. The wider the enrollment time window was opened, the more patients could enter the trial. Along the same lines, the less selective the entry criteria were in terms of comorbidities, the more likely it would be that the safety data would generalizable.

No trial design is perfect.

# Chapter 12

# FOLDING THE TENT

The Scios campus on Paseo Padre Parkway in Fremont, California, consisted of four simple rectangular buildings with tall, darkened-glass windows and rounded corners. The buildings surrounded a central plaza where, for most of the year, employees gathered for lunch outside. Toward the parking lot sat the beach-volleyball court and the dog-exercise area. (You have to love California!) Next to the buildings, a large area of as-yet-undeveloped open land ran north and east. Across the parkway to the west lay open marshland, a bird sanctuary, and then the waters of San Francisco Bay. In the morning, the parking areas were often covered in a cool, damp fog, and in the afternoon, the wind off the water carried a rich marine smell.

My first boss, Randall Kaye, left the company after the Braunwald panel events; in the crisis, I was promoted to vice president of medical affairs. It was a "battlefield promotion" if ever there was one.

I had moved into Randall's former office on the northwest corner on the second floor. It was the nicest setting in which I have ever worked. I looked out on the open land. Hawks perched along the telephone line and preyed on the ground squirrels in the field below.

Some days, you're the hawk; some days, you're the squirrel.

While the trial planning was going on, business was not going well. On June 23, 2006, J&J announced the decision to close the

Fremont office, lay off five hundred Scios and ALZA[44] employees, and move its remaining Scios workers to the ALZA facilities in Mountain View (Simmers 2006).

From the corporate financial viewpoint, the decision to downsize and relocate was sound. Natrecor sales had continued to plummet. The hoped-for Scios "pipeline" had run dry. The kinase-inhibitor compounds in development worked too well. They inhibited the kinase enzymes so effectively that they were too toxic to be successful as drugs. On the other side of the bay, after a strong start, the ALZA organization in Mountain View was having troubles of its own.

People who work in the pharmaceutical industry know that compounds in development can fail and that compounds on the market can stumble. For those with experience, job changes become part of life. If a group of pharmaceutical veterans gathers at a hotel bar at a national meeting, they will start to reminisce about the companies they worked for "back in the day." They talk about when they were on the same team and when they were competitors. It's probably the same in big-league baseball. Even so, being "impacted"—the pharma-speak euphemism for being fired—is ugly. You work with people who you come to know, like, and respect. You see the pictures of their families on their desks. All of a sudden, they're gone. Many of those remaining suffer from some degree of survivor's guilt.

As a manager, I was managing a retreat and trying to keep it orderly. Like many other pharmaceutical companies, Scios had funded a large program of investigator-initiated studies (*IIS*s). Many academic physicians, predominantly younger faculty who were trying to make a place for themselves as established clinical investigators, sent us IIS research proposals for studies related to Natrecor. Following strict internal procedures, the medical affairs group formally reviewed the proposals and funded their research on merit. Suddenly, we had a financial crisis. The IIS funds had to be reallocated to support the new safety trial. Contacting one investigator after another, I realized how critically important sound

---

44 ALZA was another Bay Area J&J operating company with its own product problems.

contracts could be. Most of the investigators were disappointed, but only a very few were angry. The angry ones made me grateful for the good contracts that our legal team had required for funding.

Scios medical affairs had also contributed very substantial funding for continuing medical education (CME) related to heart failure. Contrary to the freewheeling events that Big Pharma had funded in the late 1970s and through much of the 1980s, after J&J acquired Scios, the medical education funding process at the company complied with the strict J&J guidelines. Organizations that wanted high-quality CME programs were more than willing to work within the rules. They were required to apply for educational grants. Now those grant applications were met with the news that funds were no longer available.

Then the staff who had managed the IIS studies and the staff who had managed the CME programs took down their photos, cleared their desks, turned in their computers. Their cubicles were empty, and they were gone too.

Interestingly, for a company that was born and raised in Silicon Valley, most of the critical documents at Scios were still on paper. Soon the office spaces filled up with Iron Mountain's storage cartons. Those who were still working spent most of the day packing away the documents of two decades of Scios history. Some would go to the new office space in Mountain View; most would go into climate-controlled storage to serve out their required retention time.

* * *

The new quarters in Mountain View were in contemporary office buildings on the far eastern side of the main Google complex. Instead of a twenty-minute East Bay drive from Pleasanton, I commuted over an hour in Silicon Valley traffic. A massive glass chandelier, a Dale Chihuly[45] original, hung in the main lobby of our building. Sitting,

45 "Chihuly has led the avant-garde in the development of glass as a fine art" (Chihuly 2016).

as we did, only a few miles from the San Andreas Fault, I viewed it with great suspicion. In fact, I tried never to stand directly under it.

The excitement of developing a new drug, of bringing some relief to patients who were literally drowning in their own extracellular fluids, had given way to the personal and corporate depression of fighting a rearguard action. Morale in the remaining employees sank daily as one after another left. From the new office building, I looked out over the south end of the bay to mountains in the distance. In the foreground, a grassy rounded mound rose at the water's edge. The grass concealed a former trash dump. The Google technology geeks had designed a system for generating power for their campus from the methane gas that the trash generated on-site. Somehow, looking out on a repurposed dump seemed fitting.

* * *

The first press release for the big safety trial came on September 7, 2006. Scios announced that Duke (DCRI) would be the academic research organization leading the trial, to be called ASCEND-HF (Acute Study of Clinical Effectiveness of Nesiritide in Decompensated Heart Failure). Rob Califf would serve as the chair of the trial and would "collaborate with the Cleveland Clinic Cardiovascular Coordinating Center (C5) in managing the trial." In the press release, I was quoted as saying, "This study will join two of the largest, most experienced academic clinical research organizations in the world."

At least, I thought privately, there would be no more complaints from Steve about sloppy execution.

* * *

The media skirmishes continued without a letup. In an attempt to clean up all the Natrecor safety data that had been submitted to the FDA, the R&D team found two previously unreported deaths in patients who had received Natrecor in the PROACTION trial. On the surface, it looked bad. PROACTION was a small pilot trial with 38 sites and only 237 subjects. As it turned out, eight subjects had died within thirty days after entering the trial: one in the placebo group and seven in the Natrecor group.

As the details of the two additional deaths finally became clearer with continued investigation, the deaths were not attributable to Natrecor. One patient had left the study site against medical advice after only nine hours of treatment. He died three weeks later, while driving. The second patient died in his garage at home due to carbon monoxide inhalation. He had completed the study four weeks earlier, but he was two days shy of the thirty-day mortality reporting cutoff. Both patients were pronounced dead at hospitals not involved with the study, and the hospitals involved did not notify the study sites. Furthermore, the investigations revealed that at least two of the other Natrecor patients who were tallied as Natrecor deaths had been enrolled despite having terminal noncardiac diseases (cancer and end-stage emphysema).

PROACTION provided a textbook example of how a few unexpected and unusual events can skew the data from a small trial. That, however, did not prevent Sackner-Bernstein and Aaronson from publishing a "research letter"[46] in *JAMA* on September 27 (Sackner-Bernstein and Aaronson 2006). The letter incorporated the additional deaths into yet another highly selected set of data. The authors concluded, not surprisingly, that their analysis suggested "an association between nesiritide and increased risk of death within thirty days of its use, even at the recommended starting dose."

The *JAMA* research letter did not have a great incremental impact on its own, but it did succeed in drawing public attention away from the positive results of the NAPA trial that studied nesiritide infusion during coronary bypass surgery. The NAPA study data, which were also announced that September, showed that Natrecor infusion during coronary artery bypass surgery in patients who had impaired heart

---

46 A research letter is somewhat more formal than a letter to the editor. Research letters do not undergo the formal peer-review process required for a full scientific or medical publication.

function appeared to protect renal function and shorten hospital stay (Mentzer 2007).

No one noticed.

* * *

As 2006 ended and 2007 began, Scios had yet another discouraging development. The nine hundred–patient FUSION II study, the prospective randomized placebo-controlled study of outpatient infusions of Natrecor once or twice weekly for twelve weeks, had been the last hope for some positive clinical data to alter the downhill slide of Natrecor. But the final data from FUSION II showed no differences in either mortality or cardiovascular hospitalization rates between the subjects assigned to receive Natrecor and those who were receiving placebo infusions. The safety data with these large exposures to nesiritide were very reassuring, but there was no hint of efficacy.

The trial report was scheduled for public presentation at the annual Scientific Session of the American College of Cardiology in March of 2007. Clyde Yancy, the academic leader for trial, would present the data. He did a magnificent job. After laying out the rationale for the study and the data, he concluded, "In the context of optimal adherence to evidence based medical and device therapies, and in concert with excellent disease management, serial infusions of nesiritide did not result in a demonstrable clinical benefit over intensive outpatient management of patients with CDHF (chronically decompensated heart failure)" (Yancy 2008).

The controversy about intermittent outpatient infusions of Natrecor was now finally and completely over. They didn't work.

* * *

On June 8, 2007, the first patients were enrolled in the ASCEND-HF trial. Six years after the second advisory committee meeting, and almost two years after the Braunwald panel, Scios had embarked on a large safety trial for Natrecor. For the company, it was too little and too late.

As my medical science liaison (*MSL*) representatives and I talked with hundreds of heart failure specialists around the country, the conversations were consistent. The words would come over the telephone with a certain melancholy tone, "You know, the drug worked for my patients, and I never had any trouble with it. But after all the stuff in the *New York Times*, I just can't prescribe it. If I ever got sued, I could never defend myself."

For Scios, as for the losing boxer in a TKO, there was no single knockout punch. The company just slowly slumped to the canvas after enduring one punishing body blow after another.

The final word came on July 31, 2007. The *San Jose Mercury News* reported that "at least six hundred Bay Area employees will lose their jobs and two hundred others will be transferred to other locations because of a major cost-cutting move announced Tuesday by medical-care products giant Johnson & Johnson of New Jersey" (Johnson 2007).

* * *

By the fall of 2007, Scios was rapidly disappearing as a functional company.[47] The research and development personnel were long gone. Daniel Gennevois, a charming French infectious disease specialist, was assigned to head the J&J team for ASCEND-HF as the study responsible physician (*SRP*). But Daniel was not a Scios employee; he reported to the Janssen Research & Development operating company in Raritan. The sales force and the MSL team (my medical affairs team in the field) had all been let go—*impacted* in pharma-speak. Most of the MSLs were later rehired into J&J through the Ortho-McNeil-Janssen Pharmaceutical operating company. The same process saved jobs for many of the Scios field salespeople.

At last, Jim Mitchell called me into his office and gave me the news. I would have the opportunity to stay with J&J and to move to Ortho-McNeil-Janssen Scientific Affairs in Raritan, New Jersey. My title, as of March 2008, would be "US Natrecor Medical Affairs Lead."

---

[47] It would continue as a legal entity.

There was not much left to lead, but since J&J still had Natrecor as a product on the market, someone responsible had to mind the store.[48]

As I walked back through the empty halls, I wondered, *What will happen to the Chihuly?*[49]

---

[48] Jim Mitchell also moved back to Raritan, in an executive position with Ortho-McNeil-Janssen Pharmaceutical.

[49] Under the guidance of the J&J Office of Corporate Contributions, the 1,200-pound glass sculpture known as *Amber and Ice* was donated to UC–Davis for the main lobby of a new instructional building at the School of Veterinary Medicine.

## Chapter 13

# LIFE IN THE PENALTY BOX

> April is the cruellest month, breeding
> Lilacs out of the dead land, mixing
> Memory and desire ...
>
> —T. S. Eliot, *The Waste Land*

April 2008 fully lived up to the poet's description. According to Google maps, I had spent my three years with Scios only 2,901 miles to the west of Raritan, New Jersey. When I arrived in Raritan, I realized that I had been in a different galaxy, in a corporate culture that was light-years away.

For the past three years, I had spent every working day with people who knew every detail of the rise and fall of Natrecor. In New Jersey, no one knew what had happened at Scios; it was three time zones away. Along with the rest of the world, even the Ortho-McNeil-Janssen Scientific Affairs (OMJSA) group accepted the idea that the clanging alarms meant that there must, surely, have been a fire out there.

OMJSA occupied the third-floor rear in "the 1000 building" on the Raritan campus. When I arrived, I was assigned to a small, windowless workspace in the far front corner on the first floor. In honor of Jim Mitchell's hockey career, I nicknamed it "the penalty box." I felt as if I were in a kind of social quarantine, as if my personal stock of fleas might be carrying the black plague back from California.

After a few weeks, the rapidly growing need for a clinical cardiologist at OMJSA eroded most of the suspicions. In 2005, two very canny business development folks, who had actually been working for Scios at the time, had done a deal with the German pharmaceutical giant Bayer. J&J's Janssen pharmaceutical division would partner with Bayer to codevelop a novel anticoagulant called rivaroxaban (Caron 2005). By 2008, the development work was moving along, and OMJSA had to start planning to support an anticoagulant.

My last "new hire" at Scios was Peter Wildgoose, an experienced, energetic PhD with scientific and pharmaceutical experience in anticoagulation. When Scios imploded, Peter moved to the OMJSA organization on the rivaroxaban team. He had a window office on the third floor that he graciously let me visit. As my Natrecor activity contracted, rivaroxaban rapidly expanded to fill the gaps. But that is another story.

After six months, I moved out of the penalty box.

* * *

With a multimillion-dollar trial under way, no one in Raritan clearly understood where Natrecor or the trial fit when J&J no longer had a functioning operating company known as Scios. The solution was to form a CDT.

In Big Pharma, highly qualified people with a host of different skills work together to produce a chemically pure, sterile drug supply, to package and distribute it, to manage the required regulatory procedures, to direct research activities, and to drive the commercial process. The collection of individuals who do this for each drug is called the compound development team, or CDT. A pharmaceutical executive moving up the corporate ladder of responsibility can look forward to working as a CDT leader (*CDTL*) for two to three years, or more, to expand his or her knowledge and skills.

With the adverse publicity and commercial collapse of Natrecor, the Natrecor CDTL job was radically different from a step up the ladder—and not particularly attractive. In many ways, it was the operational equivalent of the penalty box. The few remaining internal team members supporting Natrecor and the ASCEND-HF trial breathed a collective sigh of relief when Larry Deckelbaum agreed to take on the role of Natrecor CDTL.

Larry had completed an MIT undergraduate degree and then trained in medicine and specialized in cardiology. He could be tough and intense in discussions, but his questions were always focused on the issues. He never disparaged a colleague. He was the perfect choice to pragmatically, realistically lead the team out of the confusion and depression that had settled over all of us who had experienced the collapse of Scios.

Under Larry's leadership, as the ASCEND trial moved forward, Natrecor remained on the market without a sales force, without marketing, and without advertising. The CDT covered everything related to Natrecor. Every week or so, given my role as the "scientific lead," someone in the Janssen organization would call me with new questions about a form that was missing or a database that was incomplete. There was still a Natrecor website that required regular updates. The Natrecor label also had to be updated. A manuscript or two still had to get through peer review and revision to publication. Along the way, we learned that Natrecor had been approved for use in Lichtenstein; did that mean it was on the market in the EU?[50]

Most of our thinking and planning, however, was about ASCEND-HF. Our primary academic partner, DCRI, had a truly remarkable corporate culture. Relatively junior clinical researchers were given immense operational responsibilities for large, important projects. They almost always succeeded. The approach worked, I believe, largely because the DCRI attracted extremely talented young academics. The ASCEND-HF trial fell to Adrian Hernandez at Duke. Under the leadership of Adrian and Daniel Gennevois, the trial was enrolling steadily. The CDT was responsible for developing the contingency plans for the possible outcomes. We constantly asked ourselves, "What will we do *if* ...?"

---

50 No.

Chapter 14

# NEVER FALL IN LOVE WITH A MOLECULE

The ASCEND-HF trial enrollment ran from May 2007 to August 2010. The investigators randomized 7,141 patients at 398 centers around the world. The initial results were presented in November 2010 at the American Heart Association meeting in Chicago and published in the *New England Journal of Medicine* in July 2011 (O'Connor et al. 2011).

After the AHA presentation, Larry Huston, one of the more objective reporters, wrote, "In the end everyone was wrong. After many years of controversy, we finally know the truth about nesiritide, and all the experts were wrong" (Huston 2010).

The conclusions from the *New England Journal* publication concisely summed up the findings: "Nesiritide was not associated with an increase or decrease in the rate of death and rehospitalization and had a small, nonsignificant effect on dyspnea when used in combination with other therapies. It was not associated with a worsening of renal function, but it was associated with an increase in rates of hypotension. On the basis of these results, nesiritide cannot be recommended for routine use in the broad population of patients with acute heart failure."

There was no safety problem, and there never had been. The experts, perhaps not surprisingly, did not line up to apologize for having been wrong.

On the other hand, given the trial design, the efficacy data from the trial were underwhelming. About 45 percent of patients on nesiritide and 42 percent of patients on placebo felt better after six hours of their randomized treatments. The numbers were 68 percent and 66 percent, respectively, at twenty-four hours. Those differences were statistically significant in the analysis that the European regulators required, but not in the primary analysis chosen for the United States.[51] In either case, the differences were not very impressive.

The lack of a robustly positive efficacy outcome was, at least in part, predicable from the study design. The "standard therapy" plus placebo for the control group permitted investigators to use intravenous nitroglycerin. Nesiritide and nitroglycerin have similar pharmacologic effects. As the ASCEND-HF investigators pointed out in their *New England Journal* manuscript, "The effect of nesiritide on the dyspnea end point in this trial was consistent with the findings of the Vasodilation in the Management of Acute Congestive Heart Failure trial that formed the basis for the Food and Drug Administration's approval of nesiritide." They continued, "The VMAC study included only 498 patients, and the significant effect on dyspnea at 3 hours was observed for nesiritide as compared with placebo, but this effect was similar to that of intravenous nitroglycerin, and no significant effect was detected at 24 hours."

In other words, ASCEND-HF results did not show that nesiritide was ineffective. The data did show that adding nesiritide to standard care produced outcomes that were very similar to those of a "standard care" program if the standard care included nitroglycerin—which was exactly what both the 325-326 trials and VMAC had shown earlier. In addition, nothing in the ASCEND-HF data suggested that the pulmonary pressure data from the Colucci studies or VMAC were not valid. As it turned out, the problem lay with the experts who believed that if all the pressures were better, then all the patients would undoubtedly feel better. They were wrong too.

---

[51] The issue here, for those interested, was how to allocate statistical power between the safety and efficacy end points, or "how to split the alpha." For the US analysis, the statisticians used a Bonferroni approach and allocated 95 percent of the alpha to safety. This meant that the improvement in shortness of breath would have had to be substantially greater to achieve statistical significance.

So, Huston's comment was absolutely correct. All the experts were wrong. The Scios team had focused on the efficacy comparison versus placebo, not nitroglycerin. Sackner-Bernstein and Aaronson—and Nissen—had vastly overstated the case against nesiritide. On the other hand, they had been right to insist on the importance of a large safety study.

* * *

No one left on either side had any appetite for continuing the fight.

On the company side, Scios was gone, and Natrecor sales had dwindled to almost nothing.

Among the critics, Eric Topol had departed from the Cleveland Clinic in February of 2006. As Stephanie Saul reported in the *New York Times*, Topol left "after a divisive yearlong dispute with the clinic's chief executive." She continued, "Dr. Topol said his unabashed willingness to take on Merck [over Vioxx] was at the heart of his removal" from his post as provost of the Cleveland Clinic Lerner College of Medicine (Saul 2006). When the ASCEND-HF manuscript appeared, Topol penned an accompanying editorial in the *New England Journal*; it was a comparatively mild piece that focused on his proposal that the FDA should have the authority to require post-approval studies (Topol 2011).

Steve Nissen succeeded Topol as chair of cardiovascular medicine at the clinic. Steve continued his running public commentaries on drug safety, but with his position of increased responsibility, he considerably moderated his tone.

Sackner-Bernstein continued to move steadily and rapidly through a series of jobs but published very little new or original research.

J&J had made its point; it had not marketed a dangerous drug. The final cost for ASCEND-HF was well over $100 million, but the company's reputation was intact.

Larry Deckelbaum moved on to new and greater responsibilities at J&J, a reward for his time in his own CDTL penalty box. The

Natrecor CDT was disbanded. The responsibility for Natrecor was quietly moved to the Established Products Group, and without any fanfare, advertising, or marketing, Natrecor remained available in the United States.

At Janssen, there was no enthusiasm for additional efforts in acute heart failure.

The FDA never wavered from its position. The agency had approved Natrecor for the treatment of patients with acute decompensated heart failure and dyspnea at rest or with minimal activity. The primary criterion for approval was reduction in pulmonary capillary wedge pressure. All the safety data in the "pooled analyses" had been available to the agency at the time of approval. The additional safety information from ASCEND-HF was added to the label.

Looking back, what went wrong? How did a reasonably useful, reasonably safe drug like Natrecor end up on the scrap heap? Scientists like to build theoretical models that explain their observations. The better the model, the more accurately it will predict what happens in an experiment.

If you think back to high school English, can you still recall Aristotle's definition of tragedy? He said, "Tragedy depicts the downfall of a fundamentally good person through his/her own misjudgment. The tragedy produces both suffering and insight for the protagonist and arouses the pity and fear of the audience" (Cafeharmon 2011). The Greek tragedies usually involve a character of high social status whose downfall is the product of a misjudgment. The misjudgment reflects a specific flaw, the character's inappropriate pride—or, as the Greeks called it, hubris.

From my point of view as a corporate insider, hubris at Scios and later at J&J was the root cause for the fall of Natrecor. Corporate hubris was behind the failure to move ahead rapidly with a large safety study soon after Natrecor was approved.

But tragedy is not just a story of bad luck. As the consequences of hubris unfold, they lead to a "nemesis," a catastrophic downfall. By the end, the objective is that the audience feels not only pity but also fear. That fear reflects an expanded self-awareness of human frailties. Because the structure of Greek tragedy is such a classic model of human behavior, it's also a useful model to help explain what happened to Natrecor.

The Greek tragedy analogy is a way of putting what happened to Scios and to Natrecor into a human context and of coming to understand that, after all, the characters in the drama were human. They were not bad people; they just made some very human errors. The most useful feature of putting the story into this context is that it allows us to see how some major errors in judgment and their consequences played out.

* * *

Let's look closely at what went wrong.

**The FDA**. The agency was willing to accept the Natrecor safety and efficacy data as presented in 2001, probably in part because the agency knew that the studies had been cobbled together on short timelines by a small company that had faced repeated funding crises. The data package would have looked very different if a well-heeled major pharmaceutical firm had done the clinical development work. But Scios had done it, and the company had met all the criteria that the agency had outlined. From the viewpoint of the advisory committee and the agency, the 2001 approval of Natrecor had been linked to an implied but mutually understood bargain: "Get some cash flow going, and start a large safety study!" At that time, however, the agency had no way to enforce the bargain. As a consequence, when Scios management was carried away by the heady rush of commercial success and neglected the safety study, the stage was set for future problems. This likely explains why Topol emphasized the case for giving the agency actual authority to require postmarketing studies in his "lost decade" *New England Journal* commentary (Topol 2011).

Today, the agency has the authority to require post-marketing safety studies, a major advance.

**The publications.** How did Sackner-Bernstein come up with the idea of selectively pooling subsets of the Scios trial data to highlight adverse events? And how did he involve his coauthors in the work? We will never know. The heart failure community in academic medicine is a small, close-knit group. Connections, real or imagined, between all the players are easy to make and impossible to prove. One doesn't need to channel his or her own inner Oliver Stone to invent ideas, but the idea that a simple Newtonian falling apple started the process seems unlikely.

**The press.** Press coverage of Scios following the Sackner-Bernstein publications was instrumental in shaping much of the company's fall. In her recent book, *The Bully Pulpit* (Goodwin 2013), Doris Kearns Goodwin describes the social impact of the "muckrakers," Sam McClure's band of investigative journalists that included Lincoln Steffens, Ida Tarbell, and Ray Baker, in the "golden age of journalism." Press coverage can indeed change institutions, and with such glorious predecessors, what reporter would not want to do his or her part in exposing corporate misdeeds? It's hard to fault reporters for reporting.

On the other hand, the press coverage took three very different issues and, using guilt by association, merged them into one. The issues were, first, the serious need for more data to clear up nagging questions about drug safety; second, the growing problem of physicians enthusiastically engaging in off-label use of an expensive drug; and third, the legal question of whether the company had actively promoted off-label use or had simply failed to discourage it by overlooking or even tacitly condoning the practice (the double negative is intentional). Old-fashioned in-depth reporting would have separated and clarified these issues.

**Management.** The conclusion that Scios management stumbled at multiple points along the way is impossible to avoid.

"Chapter 9. Change, and More Change" touched on the problems of organizational change, of moving from a small Bay Area biotech company to that of a national or international pharmaceutical company. The challenges of such a transition are huge. New roles, new skills, and new corporate structures are required.

A small company that has spent two decades pursuing basic new drug development without showing a profit does not have the management structure to cope with an overnight success. Dick Brewer knew this and initially tried to mitigate the situation by finding a corporate partner. Had he been successful in partnering, his approach would likely have allowed Scios to keep doing "the Bay Area biotech thing" while the partner took over the business aspects of the company.

The J&J acquisition, however, did not work that way. Scios was three thousand miles from the corporate headquarters in New Brunswick, New Jersey. In the J&J structure, Scios was acquired to become an independent operating company, but it was not ready to stand alone as a mature J&J operating company. It certainly was not ready to weather a full-blown crisis on its own.

Managers can learn from what happened to Scios. It's always tempting to say, "Let's consider the worst-case scenario." But planning for failure is pretty simple; you pack your bags and go. Planning for success, on the other hand, is very difficult. If you are heading for success, you have to fulfill your prior commitments. If you are heading for success, you *will* draw scrutiny. If you are heading for success, crises *will* happen, and you *will* need to manage them.

In *Anna Karenina*, Tolstoy wrote, "All happy families are alike; each unhappy family is unhappy in its own way." The nesiritide story is an unhappy family story, and, by Tolstoy's measure, therefore unique. A series of events that seemed superficially unrelated joined together to end in an unfortunate and unproductive outcome.

## Chapter 15

# CONSEQUENCES, FORESEEABLE AND NOT

What were the consequences of the collapse of Scios and Natrecor? These were corporate and public events with important consequences on multiple levels. One way to look at the process is to group the consequences into the personal and local fallout from the collapse of a small company and the wider systemic effects that the perceived failure of nesiritide treatment had on heart failure treatment. In addition, however, the fall of nesiritide had important, unintended consequences for the heart failure community. In my view, those unintended consequences had negative effects far more serious than were acknowledged at the time.

**Personal and Local**

First, a relatively small J&J operating company went out of business at the beginning of the economic collapse of 2008–2009. Just under a thousand people lost relatively well-paying jobs. For many, their job losses coincided with substantial losses in California's residential real estate crash. A lot of nice people became collateral damage.

**Systemic**

On a larger scale, based in part on the ASCEND-HF data, the sense of medical urgency that health care professionals had felt in managing patients with acutely decompensated heart failure seemed to erode into the concept that "standard therapy" with generic diuretics was both effective and inexpensive. The push for rapid relief of symptoms (primarily dyspnea) with vasodilator drugs had required evaluating and treating patients very soon after presentation. The choice came down to either using Natrecor, a relatively expensive drug that could be given as a fixed dose without complicated titration, or using intravenous nitroglycerin, a very inexpensive drug that requires skilled nursing and frequent dose titrations for optimal results. Somehow, the fact that ASCEND-HF showed these approaches produced roughly similar outcomes seemed to take the urgency out of the situation.

This lack of urgency may have resulted from the many health care providers who interpreted the liberal timelines for ASCEND-HF enrollment as a signal from the experts that acute heart failure was not really a medical emergency. After all, in what was supposed to be a state-of-the-art trial, the median time from a patient's presentation at the hospital to randomization for treatment in the trial was sixteen hours.

Now, five years later, rapid treatment is once again an important issue. In 2015, a new consensus document from the European Society of Cardiology Heart Failure Association (ESC-HFA), the European Society for Emergency Medicine, and the US Society for Academic Emergency Medicine (Mebazaa et al. 2015) was announced online with the headline, "Urgent Diagnosis, Treatment for Acute HF Recommended for First Time in Joint Consensus Paper" (Brauser 2015). At the 2015 meeting of the Heart Failure Society of America, both Milton Packer and Chris O'Connor advocated for early treatment (Zoler 2015) for acutely decompensated patients. By big data standards, the seven thousand–patient ASCEND-HF trial, the largest ever done in acute heart failure, is relatively small. Perhaps when we have truly robust data, we will see that the benefits of early intervention, whether with standard care or some newer agent, are real.

**Unintended**

Finally, and in my view most importantly, the law of unintended consequences came into play. Jeffery Sica, writing in *Forbes* in

2011, described the concept succinctly: "The law of unintended consequences has long existed, dating back to at least Adam Smith, but was popularized in the twentieth century by sociologist Robert K. Merton.[52] In his theory, Merton stated that *often unanticipated consequences or unforeseen consequences are outcomes that are not the outcomes intended by a purposeful action.* In some cases, the law of unintended consequences could create a perverse effect contrary to what was originally intended, and making the problem worse" (Sica 2011).

For everyone who wanted to pursue the development of new drugs for acute heart failure, and that includes pretty much everyone involved, the business outcome of the fall of nesiritide exemplified the law of unintended consequences. The characters in the nesiritide story wanted to develop safe and effective new pharmacologic treatments for patients with acute heart failure. As a result, they eventually accepted the approach that they knew would deliver the most statistically robust safety and efficacy data: the large multicenter international prospective randomized double-blind clinical trial, the ASCEND-HF trial.

The unintended consequence was that, largely in response to the public acrimony associated with Natrecor, the very smart commercial teams that drive many key decisions in large pharmaceutical companies got involved. These teams took a long, hard, and discouraging look at the economics of developing new drugs for acute heart failure. Once again, I had an insider's view of the process.

---

[52] Merton's *New York Times* obituary describes him as "one of the most influential sociologists of the 20th century, whose coinage of terms like 'self-fulfilling prophecy' and 'role models' filtered from his academic pursuits into everyday language … Mr. Merton gained his pioneering reputation as a sociologist of science, exploring how scientists behave and what it is that motivates, rewards, and intimidates them … Mr. Merton was sometimes called 'Mr. Sociology,' and Jonathan R. Cole, a former student and the provost at Columbia, once said, 'If there were a Nobel Prize in sociology, there would be no question he would have gotten it'" (Kaufman 2003).

## Chapter 16

# THE SHORT, SAD STORY OF STRESSCOPIN

Back in the "folding the tent" phase of nesiritide, as the results of the FUSION II trial were coming in, a scientist named Peter Gengo cornered me in an upstairs hallway where, for unexplained reasons, Scios kept a small collection of orchids that bloomed prolifically. Peter was one of those friends who you only meet in a small company, a laboratory guy who visited the site frequently and had lunch at the same time I did. After a couple of rambling conversations, we had established our shared enthusiasm for heart failure pharmacology.

Peter was grinning, in itself an unusual behavior given the FUSION II results. He grabbed my arm and said, "Hey, Mr. Cardiologist, have a look at this data, will you?"

He had data from a series of dog experiments that Tony Sabbah at Henry Ford, a mutual friend, had run with a new compound.

We left the orchids and headed back to my office, and after a few minutes, I said, "Peter, this looks like some kind of an inotrope."

His grin spread wider. "Yep, that's what Tony and I think too."

Over the next two years, Peter and his colleagues on the West Coast, with a kind of pharmacologic skunkworks approach, managed to scrape together funding to continue to study the new compound and move it along through the first necessary animal studies.

The molecule, called stresscopin, had been discovered in Aaron Hsueh's endocrinology laboratory at Stanford. It was another small (forty amino acids) peptide hormone. Its physiologic role was not clear, but every way Peter and Tony looked at it, it looked like an inotrope with a novel mechanism of action.

Eventually, Peter and the other scientists who were interested in stresscopin ended up clustered together in the J&J research facility in La Jolla.[53] By 2010, they had a J&J drug number for stresscopin, JNJ-39588146, shortened to *146* so everyone could remember it, and they were ready to do a preliminary phase 2 trial, "A Randomized, Double-Blind, Placebo-Controlled, Parallel-Group Study to Investigate the Safety, Tolerability, Pharmacodynamics and Pharmacokinetics of JNJ-39588146 in Subjects with Heart Failure." As with nesiritide, the preliminary trial would generate lots of raw data using the Swan-Ganz catheterization technique. Now, the Janssen R&D folks realized, they would once again need someone who could look at and interpret those data. I was asked to join a three-person oversight team for the study, and I happily agreed to take it on.

Marv Konstam's group in Boston managed data quality and coordinated the data compilation for the study, which was done at multiple sites in Europe. They did an excellent job managing the sites, and the data were encouraging. In patients with fairly severe heart failure, 146 increased cardiac output with a decrease in the resistance to forward flow, but with much smaller reductions in pulmonary wedge pressure than nesiritide and with minimal reductions in arterial blood pressure. (Hypotension, or reduction in arterial blood pressure, had been the major limitation of nesiritide.)

In 2012, the first author, Mihai Gheorghiade from Northwestern, presented the preliminary findings from the study the European Society of Cardiology Heart Failure Congress in Belgrade, Serbia. The full manuscript was published in the *European Journal of Heart Failure* in 2013 (Gheorghiade et al. 2013).

Then came the time to rally internal support. The 146 team wanted J&J to support a development program for a molecule very similar to nesiritide, a peptide hormone, for possible use in the treatment of acute heart failure. I can only describe the process as very much like John Belushi's motivational speech ("Was it over when the Germans bombed Pearl Harbor?") in *Animal House* (Belushi 1978). The fact that management's answer was a firm, even resounding, "No"

---

[53] A beautiful facility not far from the spectacular Torrey Pines golf course.

is probably not surprising, but the reasoning behind the negative response is telling.

A large clinical trial today carries a price tag of about $20,000 per randomized patient. ASCEND-HF had some seven thousand patients with a well-characterized drug. Based on the ASCEND-HF data, a projected enrollment of ten thousand for a phase 3 trial of a new acute heart failure drug with greater uncertainties about drug effect would be reasonable. In today's environment, those numbers mean a $200 million investment for the phase 3 trial *alone*. Remember that at $500 per vial, Natrecor's highly successful first year brought in just over $100 million.

Looking at the marketplace for a new drug from the commercial team's viewpoint, any single individual patient with HF might be treated on two or three occasions *at most* for acute HF during the course of his or her illness. An acute heart failure treatment, given for just a few hours, may relieve symptoms but is unlikely to improve long-term mortality. As we have seen already, that means FDA approval would have to be based on showing symptomatic improvement, the pathway that Natrecor took. But the symptomatic relief provided by a new drug would come with a very high development price tag, while in the acute HF arena today, all the "standard therapies" are inexpensive generics. Health care costs, particularly the prices of new drugs, are under intense scrutiny. When they looked at the big picture, the commercial team concluded on a purely economic basis that new drug development for acute heart failure had no viable prospects in the foreseeable future.

They were unequivocal. It wasn't just that Natrecor had collapsed, although that was a huge psychological barrier. Given the requirements of the post-Natrecor marketplace, acute heart failure drug development was over for the foreseeable future.

This was the major unintended consequence. The confrontational tactics that eventually forced J&J to fund the Natrecor safety study also killed the pharmaceutical goose that laid the golden eggs of future acute heart failure clinical research.

Natrecor had funded the ADHERE-HF registry work, the source of over three dozen publications, some of which were seminal contributions. Natrecor also funded VMAC and ASCEND-HF, which together have contributed at least thirty publications to the heart failure literature. There was not going to be any successor.

From my viewpoint, this is the real and unhappy outcome of the fall of nesiritide for both the medical community and our patients, for the Eddy Buczynskis of the world. Largely because of the huge unmet needs and the modest, but real, market opportunities, I hope and cautiously believe that the pharmaceutical industry, the academic community, and the regulatory agencies will be creative enough to devise some new, more cost-effective strategy for acute heart failure research. It cannot come soon enough.

## Chapter 17

# A NEW ORAL DRUG—NATRIURETIC PEPTIDES VINDICATED

Earlier, we discussed some of the fundamentals of cardiovascular physiologic regulation. Just to review, in patients with heart failure, impairment of cardiac pump function causes stimulation of the RAAS because of underfilling of the arterial circulation (the thermostat in the fridge). In a vicious cycle (technically, a "negative feedback system"), this RAAS overactivity causes increased resistance to forward cardiac output and retention of salt and water, putting an ever greater load on the heart. Beginning in the 1980s, large clinical trials had demonstrated that drugs that suppress RAAS activity can substantially improve outcomes in patients with heart failure and impaired pumping function. Today, those drugs, angiotensin-converting enzyme inhibitors (ACE-inhibitors) and angiotensin receptor blockers (ARBs), are standard-of-care measures in heart failure management. But modern state-of-the-art treatment can only slow the progression of heart failure. Hundreds of thousands of patients still follow in Eddy's footsteps every year.

Despite all the problems that nesiritide encountered, the idea of synergistically adding to RAAS suppression by increasing the patient's own natriuretic peptide (NP) activity still made physiologic sense. To continue the automotive metaphor, the heart failure process would slow more effectively if, in addition to taking our pharmacologic foot off the accelerator, we could put it on the brake. Nesiritide is human recombinant B-type natriuretic peptide, a polypeptide (a short form of protein). It would be digested, broken down, if taken orally, but it could be given by an injection under the skin. With the short half-life of nesiritide in the circulation, sizeable twice-daily injections for at least a couple of weeks were required to see the long-term effects. John Burnett, Horng Chen, and their colleagues at the Mayo Clinic had studied that approach with good results in a small group of patients who had heart failure and impaired pump function (Chen et al. 2012); however, administering long-term medication as multiple daily injections is costly, difficult,[54] and unpopular.

The search for a more acceptable approach to increasing NP activity led to a new idea. Instead of giving more NP, why not look for a small, orally administered molecule that could inhibit the breakdown of the patient's own (*endogenous*) NPs? Endogenous NP production is markedly increased in heart failure patients in response to increased cardiac muscle stretch. If the breakdown of the NP that the patient himself or herself was making could be slowed, then more of the hormone would be available to counteract the RAAS. There would be no need to deal with the problems of injecting a peptide drug.

To test this concept, the pathway for NP breakdown (*catabolism*) first had to be worked out, and then someone had to find a drug to interrupt that process.

Remember that in the 1990s, the Japanese investigators had established that there were, in fact, several different forms of natriuretic peptides. The two major human NPs—ANP and BNP—are cleared from the circulation by two very different mechanisms. One is a clearance receptor on the cell surface that (we believe) interlocks with the ring portion of the NP. The other is by enzymatic degradation. BNP is largely cleared by the cell surface clearance receptors that are difficult to block. However, ANP is predominantly cleared by

[54] The problems include assuring suitable temperature-controlled storage conditions, reconstituting the powdered drug with sterile liquid, and measuring the correct dose, just for starters.

a degrading enzyme called neprilysin.[55] Inhibition of neprilysin is a drugable target. This is a common phrase in the pharmaceutical industry used to indicate a technically promising mechanism of action. For those readers who are interested in the scientific details, Eugene Braunwald has reviewed the development of neprilysin inhibitors in detail (Braunwald 2015).

An orally effective neprilysin inhibitor (NEPi) called *sacubitril* was first described in 1995 (Ksander et al. 1995). Just as predicted, neprilysin inhibition with sacubitril led to increases in circulating ANP, and the increases were associated with the typical effects of ANP on cardiovascular physiology. But there was a significant downside. Sacubitril also prevented degradation of angiotensin II, a key component of the RAAS system. (One foot on the accelerator; one foot on the brake! Not so good!) Administered as a single agent, it showed limited efficacy in the treatment of heart failure or high blood pressure.

The breakthrough came when sacubitril was combined with valsartan, an angiotensin receptor blocker. In the combination, sacubitril was highly effective. The combination, called LCZ626, was invented by Webb and Ksander (Webb and Ksander 2003) and patented by Novartis in 2003. Because six molecules of valsartan and six molecules of sacubitril can actually combine in a unique crystal structure, LCZ696 was recognized as a new drug rather than just a fixed-dose combination.[56]

Based on the encouraging results from small early trials, Novartis tested LCZ696 in a large phase 3 trial led by Milton Packer and John McMurray that included 8,442 heart failure patients with impaired left ventricular ejection fraction (heart pumping function). The trial

---

[55] NPs are predominantly broken down by a membrane-bound enzyme called neprilysin, also known as neutral endopeptidase or membrane metallo-endopeptidase. Neprilysin, which is widely expressed but most abundant in the kidneys, cleaves the ring structures of the NPs, thus rendering them biologically inactive.

[56] This distinction was not only important chemically, but also critical to the approval process for LCZ626 and its subsequent marketing.

compared treatment with LCZ696 to treatment with the angiotensin-converting enzyme inhibitor, enalapril, a globally accepted standard approach for heart failure; it was discontinued early because of benefit from LCZ696. The FDA approved the drug, trade named Entresto, on July 7, 2015 (Novartis 2015).

At last, six decades after Captain Reeves started the search, persistent research around the world had finally shown a way to make the brakes work.

This does not mean that we now have a cure for heart failure, far from it. For now, all that pharmacology can do is to slow the progression of heart failure by modulating the negative feedback cycles of the disease. The ultimate goal, actually reversing the process of heart damage and normalizing the function of an injured heart, remains elusive. But so were the brakes, sixty years ago.

*Chapter 18*

# GETTING INVOLVED IN DRUG SAFETY

The up-and-down story of Scios and Natrecor has ended, and the approval of Entresto has validated the years of hard work that researchers from around the world put into understanding the structure and physiology of the natriuretic peptides.

My personal experience of engaging deeply with questions of drug safety and efficacy during my Scios years, however, helped to develop new skills and approaches to problems that had wider utility.

As Larry Deckelbaum and I worked together on the Natrecor CDT, we developed a productive professional friendship. We were both Harvard-trained cardiologists, and we shared many acquaintances through that mutual experience. We also had a lot in common in our conceptual approaches to problems. Most importantly, I was comfortable with Larry's "take no prisoners" approach to technical discussions.

Larry and Kate Cabot, another J&J cardiologist, had jointly established the J&J Cardiovascular Safety Group (CVSG) to formalize the process of cardiac safety consultation for compound development teams (CDTs) that wanted expert advice. There were relatively few cardiologists at J&J, and in 2009, not long after I had moved out of the penalty box, Larry and Kate asked me to join the CVSG. I agreed, with little understanding of how very much I had to learn.

I had a good practical working understanding of the importance of maintaining a stable cardiac rhythm based on my clinical experience in coronary care, in pacemaker implantation, and in caring for patients with advanced heart failure. I knew far less about the sciences of electrophysiology, pharmacology, and epidemiology.

Cardiac safety programs in the pharmaceutical industry first grew out of a pragmatic need to design studies and interpret electrophysiological data, including the findings of what are now called *thorough QT studies*, in order to assess the arrhythmia risk associated with new drugs in development. Initially, the cardiac safety strategy was fairly limited: either terminate development of a new drug with an apparent QT liability before incurring the expense of large clinical trials or develop risk-mitigation strategies for agents with great potential benefit.

Soon, however, other drug-related cardiac risks in addition to arrhythmias began to surface. Chemotherapy for cancer has steadily become more effective. As patients enjoy more prolonged remissions or perhaps even cures, some have also developed chemotherapy-related heart failure. Several anticancer drugs can cause long-term heart muscle damage with eventual heart failure as an off-target effect.

New and better drugs for the more severe forms of mental illness, particularly schizophrenia and treatment-resistant depression, have also recently come into clinical development. Some of these drugs have important effects on heart rate and blood pressure. Because these adverse effects target the heart and blood vessels, the compound development teams (CDTs) started to ask their cardiovascular safety colleagues about these new problems as well.

In theory, the strategy of just discontinuing development for drugs that have a cardiovascular risk sounds both rational and efficient. In practice, it is incredibly difficult. Often difficulties arise because there are no objective standards to define the magnitude of potential benefit. Here, drug safety must look to the epidemiologists for help. A potent new chemotherapy for a particular cancer or an effective antimicrobial for serious drug-resistant infections might have great potential benefit for a very limited number of patients. In that situation, the potential benefit might justify accepting substantial risk for individuals who would otherwise be likely to die without treatment.

In contrast, essentially no potential for serious risk would be acceptable for a new agent for high blood pressure or high cholesterol that might be given to hundreds of thousands of patients. We already have safe and effective alternative drugs for those patients, and the magnitude of exposure would guarantee that some patients would experience adverse effects. For example, the recall of Baycol (cerevastatin) involved thirty-one reported deaths in the United States out of over seven hundred thousand patients estimated to have been exposed, but safe and effective alternatives were available, with five other effective statin drugs on the market (CBS News 2006).

The cardiovascular safety group membership included representatives from a wide spectrum of disciplines, including laboratory and clinical pharmacology, epidemiology, biostatistics, regulatory affairs, and the various therapeutic development areas, as well cardiology. They wrestled with these complex risk-benefit questions as consultants for multiple CDTs.

* * *

In 2011, as the ASCEND-HF trial came to a close, Larry Deckelbaum prepared to move on from his position as the Natrecor CDT leader, and Kate Cabot announced her retirement. Neither Kate nor Larry would be able to continue to lead the J&J Cardiovascular Safety Group. The committee was a J&J "Center of Excellence" and its primary support came from the cardiovascular group in research and development. Even though at that time, I was a commercial—not an R&D—employee, Larry asked me to step in as the acting chair of the CVSG. Both of them felt that the chair of the committee should be a cardiologist with clinical experience. As a former medical school professor, I had wanted to understand the drug safety issues more completely. What better way to learn than to teach the course? I agreed to take it on for a few months, until a suitable successor could be identified. Then, in 2011, I moved from Janssen Scientific Affairs (commercial) to Janssen Research & Development. I turned out to meet the requirement for the chair position as an R&D employee and became the "suitable successor."

Over the next five years, I chaired some three dozen formal CVSG meetings. With the depth of experience and talent that the other CVSG members brought to the table, explaining the cardiac aspects of the various safety issues that came to the team was straightforward. Increasingly evident was that the complexity lay in the epidemiology and biostatistics aspects of the problems we discussed. These areas were where having experience with the Natrecor safety controversies of the past few years made an important impact. Understanding the potential for unusual events to occur in small studies, the need for large numbers to support statistical analyses, and the importance of having clinical detail in adverse event reports helped me keep my composure when some clinical teams were prepared to push the panic button.

Building on the Natrecor experience, over the course of chairing the CVSG meetings, I began to realize that the most important role for the drug safety group was to help develop good questions rather than to provide tentative answers.

Perhaps just one short concrete example will help to clarify the issue.

The default assumption regarding the cardiac safety of antipsychotic drugs is that persons with severe mental illness are healthy except for their disturbances of thought and emotion. In fact, nothing could be further from the truth. Strong epidemiologic data show that schizophrenic or severely depressed patients do not live a healthy lifestyle. These patients may be homeless. They often smoke, and many engage in substance abuse. They do not follow healthy balanced diets; they don't get regular medical and dental care, and if they also happen to have diabetes or high blood pressure, they don't take the medications required to control those conditions. All of these factors combine to significantly increase the risk of heart disease for schizophrenic patients (Laursen 2012).

Let's hypothesize that the CDT for a new, long-acting antipsychotic agent finds that it may have a low, but real, potential for adverse cardiac effects. On the other hand, this new drug measurably improves functioning of a significant number of schizophrenic patients. What information will be needed to understand the overall risk-benefit proposition for such a drug? Will the overall rate of cardiac events in the patients who receive the drug (the *target patient population*) increase because of the added risk of the drug? Or will cardiac events decrease because along with their improved psychiatric state, the target patients would also follow a healthier lifestyle, have better diets, and manage their diabetes and high blood pressure more effectively?

By understanding the epidemiology and the statistical problems, a drug safety team can help get to the critical questions. The drug development team can then work both backward, reanalyzing their existing data, and forward, building in acquisition of the new data required as they write their new protocols.

Two or three decades ago, making even rough estimates in an attempt to answer those questions would have been impossible. Today, with the large health databases available to professional epidemiologists, and with the powerful computational tools in the hands of biostatisticians, modeling the problems with different inputs is possible. Not every problem requires the time, effort, and expense of developing quantitative models, but some do. As a result, many times the discussion in a cardiovascular safety meeting becomes "Is this a question that we really must answer? If the drug's CDT actually had quantitative answers to these questions, would those answers help us decide whether or not to continue to develop the drug? And would they help us to plan how to do post-marketing safety surveillance?"

In summary, first my work with nesiritide and then my in-depth experience with cardiovascular drug safety taught me that sounding an alarm without knowing whether it's true or false is a simple gut reaction. It's the first thing any of us want to do when we sense danger. In contrast, developing a thoughtful, factual response to a potential problem and generating the data to answer the problem takes discipline, time, a coordinated effort, and resources. An appropriate response to a safety signal asks questions: What is the science, what are the data, and what do the epidemiology and biostatistics specialists think?

## Chapter 19

# RISK-BENEFIT—ANOTHER FDA ADVISORY COMMITTEE

In 2011, as a member of a small risk-benefit team, I helped with the Janssen R&D preparations for another FDA advisory committee meeting. This time, we were dealing with a new drug, rivaroxaban, that had been tested in a fourteen thousand–patient trial of stroke prevention in patients with atrial fibrillation. Comparing the data and the effort that went into the risk-benefit and safety discussion in 2011 to the safety issues as Scios approached them ten years earlier in 2001 shows why the post-approval rise and fall of nesiritide is unlikely to happen again with a new drug. As you have already seen, the future of a potential new drug hangs in the balance at an FDA advisory committee meeting. Presenting the scientific evidence that supports the drug clearly and concisely is the key to success. Both Scios and the Janssen R&D team knew that, so it's instructive to compare the two events.

With both nesiritide and rivaroxaban, the major safety problem is a direct result of the mode of action.[57] Basically, the drug does what it is supposed to do but does it to excess. Nesiritide causes blood vessel dilation; the major safety issue with nesiritide is hypotension (low blood pressure). Rivaroxaban inhibits an

[57] "An important strategy for assessing safety is to carefully consider the likely adverse consequences of a product's desired clinical effects" (Friedhoff 2009).

important step in the clotting system; the major safety issue with rivaroxaban is bleeding.

The nesiritide safety data included a total of fewer than one thousand patients aggregated from nine different studies. The studies had used different inclusion and exclusion criteria, had exposed patients to different doses of the drug for different periods, and the safety follow-up had not been managed using standardized measures. In contrast, just a decade later, the single phase 3 ROCKET trial had randomized over fourteen thousand patients, and over seven thousand of them had received rivaroxaban. They were enrolled, managed, and followed under a uniform protocol.

The magnitude of change reflects the incredible impact that digital technology has had on clinical research. Patients are enrolled and followed in trials using electronic case report forms. Laboratory data are transmitted to a central repository using reliable data-transfer technology. The computational aspects of sophisticated statistical analyses can be performed quickly and reliably.

When nesiritide was approved and marketed as Natrecor, Sackner-Bernstein and his colleagues could use various criteria to pick and choose among the different studies. From these smaller pools of patients, they were able to generate skewed safety data and then raise what proved to be a false alarm based on retrospective ad hoc subgroup analyses. In contrast, the rivaroxaban data from a single large study were uniform and robust.

The rigor with which patient-reported outcomes and patient preferences were studied provides another striking contrast between the preparations for the two advisory boards. In the VMAC and ASCEND studies, patients were simply asked whether their dyspnea was better, unchanged, or worse at various times after treatment. The patients assigned a value on a seven-point scale (the Likert scale) to how they felt. This approach measuring change in dyspnea (technically, a patient-reported outcome) has never been objectively validated and was never formally correlated with other clinical and demographic data, including the changes in wedge pressure. It's not difficult to imagine the multiple problems involved. The subjective changes that a patient experiences must, in some way, be related to the cause and severity of his or her heart disease, how rapidly the symptoms developed, how long they have been present, and how severe they were, to say nothing of the presence or absence of lung disease as well. The seven-point scale alone cannot capture any of this subtlety.

In contrast, in preparation for the rivaroxaban advisory committee, the risk-benefit group conducted extensive and carefully structured interviews with patients and physicians to explore their attitudes toward various bleeding events and toward stroke events from mild to disabling.

The difference in approach, in just ten years, is exciting and encouraging. I like to think that, in part, the changes reflect lessons learned from the nesiritide experience.

Finally, the FDA now has the authority to require post-marketing safety studies. This important and positive step combines three important concepts and developments. First, drug safety events are fortunately rare, and because they are rare, they don't occur in noticeable numbers until large numbers of patients have been exposed. Second, with electronic health and health-insurance data, it's now possible to identify and observe very large numbers of patients exposed to a new drug. Third, modern data transfer and analysis makes it possible to analyze all this observational data and to apply the statistical corrections for multiple variables that make it interpretable.

Increased attention to post-marketing safety is a logical consequence of attempts to speed up the drug approval process. The FDA website succinctly describes the process. "On September 27, 2007, the President signed the Food and Drug Administration Amendments Act of 2007 (FDAAA) (Public Law 110-85). Section 901, in Title IX of FDAAA, created section 505(o) of the Federal Food, Drug, and Cosmetic Act (the Act), which authorizes FDA to require certain studies and clinical trials for prescription drugs and biological products approved under section 505 of the Act or section 351 of the Public Health Service Act. This new authority became effective on March 25, 2008. Postmarketing studies or clinical trials may be required to

- assess a known serious risk related to the use of a drug,
- assess signals of serious risk related to the use of the drug, and
- identify an unexpected serious risk when available data indicate the potential for a serious risk (FDA, Home Drugs Guidance, Compliance & Regulatory Information Postmarket Requirements and Commitments, n.d.).

At the end of the day, the answer to "How much drug-related risk is acceptable?" must always be "Compared to what benefit?" Understanding how people view risk and reward is neither simple nor intuitive (Kahneman 2011). Developing objective, rigorous methods to understand the expected value of the trade-offs involved with new therapeutic agents is a new frontier for clinical trialists and a rich field for continuing research.

Returning for a moment to lessons learned from nesiritide, just consider: what if the trials in the clinical development program for nesiritide had used uniform safety reporting criteria and standardized follow-up? What if Scios had had the money, time, and expertise to build and validate a better patient-reported outcome instrument? And what if Scios had carefully studied how much the patients actually valued earlier relief from their labored breathing and how much cost saving came with the simplicity of using nesiritide?

Of course, these questions are what-ifs.

In telling this story, I hope that I have clearly shown the importance of sound process. For nesiritide, the major problems related to management decisions and drug safety data. The lesson learned is that sound process includes being as relentless, unemotional, and rational as Joe Friday. To make good decisions, both managers and scientists must gather as much data as are needed, objectively analyze the data with multidisciplinary approaches, and develop appropriate questions to investigate and resolve the problems they face. It is a generalizable approach. It is the way a liberal education teaches us to approach problems. It is not the easy way or the popular way.

The steady increase in the tempo of our electronic communication tempts us to sacrifice thoughtful attempts to gather facts and reflect on issues for the satisfaction of an immediate response. Across multiple disciplines in science, medicine, education, and public affairs, we deal with enormously complicated issues. *Nesiritide: The Rise and Fall of Scios* is a cautionary tale on many levels, but altogether it makes a strong case for process.

As Joe said, "All we want are the facts, ma'am" (Snopes.com 2008).

# *Afterword*

When I started out to write *Nesiritide*, I had a title and a detailed outline. I had no plans for an afterword and did not see the need for one. Having come to the end of the writing, after reliving, researching, and reflecting on the events, I want to add a personal perspective.

A careful review of the source documents that I have cited in the text has convinced me that the individuals involved in the fall of Scios were not bad people. Those who demanded a large safety study fervently believed in their cause. It should have been the top priority at Scios after approval. Failure to heed the obvious warnings carried a heavy price tag.

One of the goals of tragic drama is to instill fear in the audience, the fear of making the same mistake.

# *References*


Adams, D., and M. Carwardine. 1991. *Last Chance to See.* New York: Harmony Books.

Baines, A. D., A. J. de Bold, and H. Sonnenberg. 1983."Natriuretic effect of atrial extract on isolated perfused rat kidney." *Canadian Journal of Physiology Pharmacology* 61 (12): 1462–1466.

Belushi, John. 1998. "Bluto's Speech from *Animal House.*" YouTube video, 2:45. From movie released June 1, 1978. Posted by "Michael Kirkpatrick," October 28 2012. Accessed February 22, 2016. https://www.youtube.com/watch?v=ep-xgd_eETE.

Binder, Gordon, and Philip Bashe. 2008. *Science Lessons: What the Business of Biotech Taught Me about Management.* Boston: Harvard Business Press.

Braunwald, E. B. 2015. "The path to an angiotensin receptor—neprilysn inhibitor in the treatment of heart failure." *Journal of the American College of Cardiology* 65 (10): 1025–41.

———. 2005. "Expert Panel Recommendations." June 13 press release. Fremont, CA: Scios.

———. 2004. *Heart Disease: A Textbook of Cardiovascular Medicine.* 6th ed. Philadelphia: Elsevier Health Sciences.

Brauser, Deborah. 2015. "The Heart.org." June 4. Accessed November 20, 2015. http://www.medscape.com/viewarticle/845913.

Burger, A. J., U. Elkayam, M. T. Neibaur, H. Haught, J. Ghali, D. P. Horton, and D. Aronson. 2001. "Comparison of the occurrence of ventricular arrhythmias in patients with acutely decompensated congestive heart failure receiving dobutamine versus nesiritide therapy." *American Journal of Cardiology* 88 (1): 35–39.

Cafeharmon [username]. 2011. "Elements of the Greek Tragedy and the Greek Hero." April 11. Accessed October 2015. http://www.slideshare.net/cafeharmon/elements-of-greek-tragedy-and-the-tragic-hero.

Caron, Nancy. 2005. "Bayer, Johnson & Johnson Sign Collaborative Agreement for Antithrombosis Drug." *First Word Phama*. October 26. Accessed November 20, 2015. www.firstwordpharma.com/node/177848#axzz3n8e2SF83.

CBS News. 2006. "Deaths Spur Cholesterol Drug Recall." July 10. Accessed November 20, 2015. http://www.cbsnews.com/news/deaths-spur-cholesterol-drug-recall-10-07-2006/.

Chen, H. H., J. F. Glockner, J. A. Schirger, A. Cataliotti, M. M. Redfield, and J. C. Burnett Jr. 2012. "Novel protein therapeutics for systolic heart failure: Chronic subcutaneous B-type natriuretic peptide." *Journal of the American College of Cardiology* 60 (22): 2305–2312.

Chihuly. 2016. "Learn." Accessed May 2. http://www.chihuly.com/learn.

Cleveland Clinic. 2016. "Cleveland Clinic Lerner College of Medicine of Case Western Reserve University." Accessed January 5. http://portals.clevelandclinic.org/cclcm/CollegeHome/tabid/7343/Default.aspx.

Cleveland Clinic. 2015. "Heart and Vascular Institue." Accessed October 26. http://my.clevelandclinic.org/services/heart?utm_camp

aign=national+usnews+ads&utm_medium=banner&utm_source=usnews&utm_content=pb+300x250+ccmain+21_1hrt&dynid=usnews-_-national+usnews+ads-_-banner-_-pb+300x250+ccmain+21_1hrt-_-pb+300x250+ccmain+21_1hrt.

Colucci, W. S., U. Elkayam, D. P. Horton, W. T. Abraham, R. C. Bourge, A. D. Johnson, L. E. Wagoner, M. M. Givertz, C. S. Liang, M. Neibaur, W. H. Haught, and T. H. LeJemtdel. 2000. "Intravenous nesiritide, a natriuretic peptide, in the treatment of decompensated congestive heart failure." *New England Journal of Medicine* 343 (4): 246–253.

de Bold, A. J. 1985. "Atrial natriuretic factor: A hormone produced by the heart." *Science* 230 (4727): 767–770.

Dexter, L., F. W. Haynes, et al. 1947."Studies of congenital heart disease: The pressure and oxygen content of blood in the right auricle, right ventricle, and pulmonary artery in control patients, with observations on the oxygen saturation and source of pulmonary capillary blood." *Journal of Clinical Investigation* 26 (3): 554–560.

Dikshit K, J.K. Vyden, J.S. Forrester, K. Chatterjee, R. Prakash, H.J. Swan. 1973. "Renal and extrarenal hemodynamic effects of furosemide in congestive heart failure after acute myocardial infarction." *New England Journal of Medicine* 288 (21): 1087–1090.

Ellenberg, J. 2014. *How Not to Be Wrong: The Power of Mathematical Thinking.* New York: Penguin Books.

Forrester, James S. 2015. *The Heart Healers.* New York: St. Martin's Press.

Friedhoff, L. T. 2009. *New Drugs: An Insider's Guide to the FDA's New Drug Approval Process.* New York: Pharmaceutical Special Projects Group.

Gheorghiade, M., S. J. Greene, P. Ponikowski, et al. 2013. "Haemodynamic effects, safety, and pharmacokinetics of human stresscopin in heart failure with reduced ejection fraction." *European Journal of Heart Failure* 15 (6): 679–689.

Gheorghiade, M., C. Orlandi, J. C. Burnett Jr., et al. 2005. "Rationale and design of the multicenter, randomized, double-blind, placebo-controlled study to evaluate the efficacy of vasopressin antagonism in heart failure: Outcome study with tovaptan (EVEREST)." *Journal of Cardiac Failure* 11 (4): 260–269.

Goodwin, Doris Kearns. 2013. *The Bully Pulpit: Theodore Roosevelt, William Howard Taft, and the Golden Age of Journalism.* New York: Simon & Schuster.

Habak, P. A., A. L. Mark, J. M. Kioschos, et al. 1974. "Effectiveness of congesting cuffs ('rotating tourniquests') in patients with left heart failure." *Circulation* 50 (2): 366–371.

Henry, J. P., and J. W. Pearce. 1956. "The possible role of cardiac atrial stretch receptors in the induction of changes in urine flow." *Journal of Physiology* 131 (3): 572–585.

Hughes, Sally Smith. 2011. *Genentech: The Beginnings of Biotech.* Chicago: University of Chicago Press.

Huston, Larry. 2010. *Cardiobrief.* November 14. Accessed October 2015. cardiobrief.org/2010/11/14/aha-ascend-trial-finally-resolves-the-nesiritide-controversy/.

Johnson & Johnson. 2016. "Our Credo Values." Accessed January 4. http://www.jnj.com/about-jnj/jnj-credo.

Johnson, Steve. 2007. "Johnson & Johnson Cuts Bay Area Divisions." *San Jose Mercury News*, July 31.

Kahneman, Daniel. 2011. *Thinking, Fast and Slow.* New York: Farrar, Straus and Giroux.

Kambayashi, Y., K. Nakao, M. Mukoyama, Y. Saito, Y. Ogawa, S. Shiono, K. Inouye, N. Yoshida, and H. Imura. 1990. "Isolation and sequence determination of human brain natriuretic peptide in human atrium." *FEBS Letters* 259 (2): 341–345.

Kaufman, Michael T. 2003. "Robert K. Merton, Versatile Sociologist and Father of the Focus Group, Dies at 92." *New York Times*, February 24. http://www.nytimes.com/2003/02/24/nyregion/robert-k-merton-versatile-sociologist-and-father-of-the-focus-group-dies-at-92.html?pagewanted=all.

Kipling, Rudyard. 2015. "If—" Poetry Foundation. Accessed October. www.poetryfoundation.org/poem/175772.

Ksander, G. M., R. D. Ghai, R. deJesus, et al. 1995. "Dicarboxylic acid dipeptide neutral endopeptidase inhibitors." *Journal of Medicinal Chemistry* 38 (10): 1689–1700.

Laursen, T. M., T. Munk-Olsen, and M. Vestergaard. 2012. "Life expectancy and cardiovascular mortality in persons with schizophrenia." *Current Opinion in Psychiatry* 25 (2): 83–88.

Leuty, Ron. 2012. "Longtime Biotech Exec Richard Brewer Dies from Multiple Myeloma." *San Francisco Business Times*, August 14. Accessed December 14, 2015. http://www.bizjournals.com/sanfrancisco/blog/biotech/2012/08/richard-dick-brewer-dendreon-nile-died.html.

Lown, B., and H. D. Levine. 1958. *Atrial Arrhythmias, Digitalis, and Potassium.* New York: Landsberger Medical Books.

Marcus, L. S., D. Hart, M. Packer, M. Yushak, N. Medina, R. S. Danziger, D. F. Heitjan, and S. D. Katz. 1996. "Hemodynamic and renal excretory effects of human brain natriuretic peptide infusion in patients with congestive heart failure. A double blind placebo controlled randomized crossover trial." *Circulation* 94 (12): 3184–3189.

Maugh, Thomas H. 2009. "John F. Kennedy's Addison's disease was probably caused by rare autoimmune disease" *LA Times*, September 5. http://articles.latimes.com/2009/sep/05/science/sci-jfk-addisons5.

Mebazaa, A., B. Yimaz, P. Levy, et al. 2015. "Recommendations on pre-hospital and early hospital management of acute heart failure." *European Heart Journal*: 1958–1966.

Mentzer Jr., R. M., M. C. Oz, R. N. Sladen, et al. 2007. "Effects of perioperative nesiritide in patients with left ventricular dysfunction undergoing cardiac surgery: The NAPA Trial." *Journal of the American College of Cardiology* 49 (6): 716–726.

Mills, R. M. 2009. "The heart failure frequent flyer: An urban legend." *Clinical Cardiology* 32 (2): 67–68.

———. 1971. "Severe hypersensitivity reactions associated with allopurinol." *Journal of the American Medical Association* 216 (5): 799–802.

*New York Times*. 2003. "Johnson & Johnson to Buy Scios for $2.4 billion." February 11. http://www.nytimes.com/2003/02/11/business/company-news-johnson-johnson-to-buy-scios-for-2.4-billion.html.

Novartis. 2015. "Novartis' New Heart Failure Medicine LCZ696, Now Called Entresto(TM), Approved by FDA to Reduce Risk of Cardiovascular Death and Heart Failure Hospitalization." July 7. Accessed November 20, 2015. https://www.novartis.com/news/media-releases/

novartis-new-heart-failure-medicine-lcz696-now-called-entrestotm-approved-fda.

O'Connor, C. M., R. C. Starling, A. F. Hernandez, et al. 2011. "Effect of nesiritide in patients with acute decompensated heart failure." *New England Journal of Medicine* 365 (1): 32–43.

Ohio State University Wexner Medical Center. 2016. "William Abraham, MD." Accessed May 2. http://wexnermedical.osu.edu/patient-care/find-a-doctor/william-abraham-md-19621.

Packer, M., W. Colucci, L. Fisher, et al. 2013. "Effect of levosimendan on the short-term clinical course of patients with acutely decompensated heart failure." *Journal of the American College of Cardiology* 1 (2): 103–111.

Patel, M. R., K. W. Mahaffey, J. Garg, et al. 2011. "Rivaroxaban versus warfarin in non-valvular atrial fibrillation." *New England Journal of Medicine* 365 (10): 883–891.

Perzborn, E., S. Roehrig, A. Straub, D. Kubitza, and F. Misselwitz. 2011. "The discovery and development of rivaroxaban, an oral, direct factor Xa inhibitor." *Nature Reviews, Drug Discovery* 10 (1): 61–75.

Pettica, Mike. 2013. "'Real' Doubleheaders." *Cleveland.com.* May 12. http://www.cleveland.com/tribe/index.ssf/2013/05/real_doubleheaders_were_once_h.html.

*Pink Sheet.* 1999. May 3.

Publication Committee for VMAC. 2002. "Intravenous nesiritide vs nitroglycerin for treatment of decompensated congestive heart failure: A randomized controlled trial." *Journal of the American Medical Association* 287 (12): 1531–1540.

Sackner-Bernstein, J., and K. Aaronson. 2006. "Research Letter." *Journal of the American Medical Association* 296: 1465–1466.

Sackner-Bernstein, J., M. Kowalski, M. Fox, and K. Aaronson. 2005. "Short-term risk of death after treatment with nesiritide for decompensated heart failure: A pooled analysis of randomized controlled trials." *Journal of the American Medical Association* 293 (15): 1900–1905.

Sackner-Bernstein, J., H. A. Skopicki, K. D. Aaronson. 2005. "Risk of worsening renal function with nesiritide in patients with acutely decompensated heart failure." *Circulation* 111 (12): 1487–1491.

Saul, Stephanie. 2006. "High-Profile Doctor to Leave Position at Cleveland Clinic." *New York Times*, February 10.

———. 2005a. "Heart Clinic May End or Curtail Use of Drug." *New York Times*, May 4. http://www.nytimes.com/2005/05/04/health/heart-clinic-may-end-or-cutail-use-of-a-drug.

———. 2005b. "Johnson & Johnson Adds Data on Deaths to Label on Heart Treatment." *New York Times*, April 26.

———. 2005c. "The Marketing and Success of Natrecor." *New York Times*, May 17. http://www.nytimes.com/2005/05/17business/the-marketing-and-success-of-natrecor.html.

———. 2005d. "Popular Heart Drug Impairs Kidney Action, Study Says." *New York Times*, March 22.

———. 2005e. "U.S. Looking at Marketing by Johnson & Johnson." *New York Times*, July 21. http://www.nytimes.com/2005/07/21/business/us-looking-at-marketing-by-johnson-johnson.html.

Scios. 2005a. "FUSION I." December 22. https://clinicaltrials.gov/ct2/show/NCT00270361.

Scios. 2005b. "VMAC." December 22. Accessed October 22, 2015. https://clinicaltrials.gov/ct2/show/NCT00270374?term=VMAC.+Vasodilation+in+the+management+of+acute+congestive+heart+failure&rank=1, 2000.

Scios. 2003. February 13 press release. Fremont, CA: Scios.

Sheldon, William C. 2008. *Pathfinders of the Heart: The History of Cardiology at the Cleveland Clinic.* Bloomington, IN: Xlibris.

Sica, Jeffery. 2011. "The Law of Unintended Consequences: The Worst Mistake in Decades." *Forbes*, February 28.

Simmers, Tim. 2006. *BioSpace.* June 23. Accessed October 2015. http://www.biospace.com/Default.aspx.

Singer, D. E., Y. Chang, M. C. Fang, L. H. Borowsky, N. K. Pomernacki, N. Udaltsova, and A. S. Go. 2009. "The net clinical benefit of warfarin anticoagulation in atrial fibrillation." *Annals of Internal Medicine* 151 (5): 297–305.

Snopes.com. 2008. "*Dragnet*." December 13. Accessed November 23, 2015. http://www.snopes.com/radiotv/tv/dragnet.asp.

*This Day in Quotes.* 2015. February 7. http://www.thisdayinquotes.com/2010/02/it-became-necessary-to-destroy-town-to.html.

Topol, E. J. 2011. "The lost decade of nesiritide." *New England Journal of Medicine* 365 (1): 81–82.

———. 2005. "Nesiritide—not verified." *New England Journal of Medicine* 353 (2): 113–116.

———. 2004. "Failing the public health—Rofecoxib, Merck, and the FDA." *New England Journal of Medicine* 351 (17): 1707–1709. doi: 10.1056/NEJMp048286.

———. 2002. *Textbook of Cardiovascular Medicine.* Philadelphia: Lippincott Williams & Wilkins.

UPI. 2005. "Heart Drug May Increase Risk of Death." April 19. Accessed October 2015. http://www.upi.com/Science_News/2005/04/19/Heart-drug-may-increase-risk-of-death/31591113941093/.

US FDA (US Food and Drug Administration). 2015a. "History." Last updated March 23. http://www.fda.gov/AboutFDA/WhatWeDo/History/.

———. 2015b. "Home Drugs Guidance, Compliance & Regulatory Information Postmarket Requirements and Commitments." Last updated December 21. Accessed November 20, 2015. http://www.fda.gov/Drugs/GuidanceComplianceRegulatoryInformation/Post-marketingPhaseIVCommitments/ucm070766.htm.

———. 2015c. "Press Announcements." January 26. http://www.fda.gov/NewsEvents/Newsroom/PressAnnouncements/ucm431571.htm.

———. 2011. "FDA Press Announcements." July 5. Accessed November 20, 2015. www.fda.gov/NewsEvents/Newsroom/PressAnnouncements/ucm261839.htm.

———. 2001. "United States of America, Department of Health and Human Services, Food and Drug Administration, Center for Drug Evaluation and Research, Cardiovascular and Renal Drugs Advisory Committee 92nd Meeting." May 25. http://www.fda.gov/ohrms/dockets/ac/01/transcripts/3749t2_01.pdf.

———. 1999. "US FDA CDER Cardiovascular and Renal Drugs Advisory Committee 87th Meeting." January 29. Accessed December 14, 2015. http://www.fda.gov/ohrms/dockets/ac/99/transcpt/3490t2.pdf.

Wardrop, D., and D. Keeling. 2008. "The story of the discovery of heparin and warfarin." *British Journal of Haematology* 141 (6): 757–763.

Watson, J. D., and F. H. C. Crick. 1953. "A structure for deoxyribose nucleic acid." *Nature* 171: 737–738.

Webb, R. L., and G. M. Ksander. 2003. Pharmaceutical compositions comprising valsartan and NEP inhibitors. International Patent Application PCT/EP03/00415.

*Wikipedia.* 2016. "Joe Friday." Last updated February 9. Accessed January 5, 2016. https://en.wikipedia.org/wiki/Joe_Friday.

Winchester, Simon. 2010. *Atlantic.* New York: HarperCollins.

Yancy, C. W., H. Krum, B. M. Massie, et al. 2008. "Safety and efficacy of outpatient nesiritide in patients with advanced heart failure: Results of the Second Follow-Up Serial Infusions of Nesiritide (FUSION II) trial." *Circulation Heart Failure* 1 (1): 9–16.

Yancy, C. W., M. T. Saltzberg, R. L. Berkowitz, B. Bertolet, K. Vijayaraghavan, K. Burnham, R. M. Oren, K. Walker, D. P. Horton, and M. A. Silver. 2004. "Safety and feasibility of using serial infusions of nesiritide for heart failure in an outpatient setting (from the FUSION I trial)." *American Journal of Cardiology* 94 (5): 595–601.

Yuan, Z., B. Levitan, P. Burton, et al. 2014. "Relative importance of benefits and risks associated with antithrombotic therapies for acute coronary syndrome: Patient and physician perspectives." *Current Medical Research and Opinion* 30 (9): 1733–1741.

Zoler, Mitchel L. 2015. "Emphasizing 'acute' in ADHF treatment." *Cardiology News* (December): 1, 24–25.

# TRUE DIRECTIONS

*An affiliate of Tarcher Perigee*

## OUR MISSION

Tarcher Perigee's mission has always been to publish books that contain great ideas. Why? Because:

## GREAT LIVES BEGIN WITH GREAT IDEAS

At Tarcher Perigee, we recognize that many talented authors, speakers, educators, and thought-leaders share this mission and deserve to be published – many more than Tarcher Perigee can reasonably publish ourselves. True Directions is ideal for authors and books that increase awareness, raise consciousness, and inspire others to live their ideals and passions.

Like Tarcher Perigee, True Directions books are designed to do three things: inspire, inform, and motivate.

Thus, True Directions is an ideal way for these important voices to bring their messages of hope, healing, and help to the world.

Every book published by True Directions– whether it is non-fiction, memoir, novel, poetry or children's book – continues Tarcher Perigee's mission to publish works that bring positive change in the world. We invite you to join our mission.

For more information, see the True Directions website:

www.iUniverse.com/TrueDirections/SignUp

Be a part of Tarcher Perigee's community to bring positive change in this world! See exclusive author videos, discover new and exciting books, learn about upcoming events, connect with author blogs and websites, and more! www.tarcherbooks.com

Printed in the United States
By Bookmasters